T0213227

SpringerBriefs in Computer Science

Series editors

Stan Zdonik, Brown University, Providence, USA
Shashi Shekhar, University of Minnesota, Minneapolis, USA
Jonathan Katz, University of Maryland, College Park, USA
Xindong Wu, University of Vermont, Burlington, USA
Lakhmi C. Jain, University of South Australia, Adelaide, Australia
David Padua, University of Illinois Urbana-Champaign, Urbana, USA
Xuemin (Sherman) Shen, University of Waterloo, Waterloo, Canada
Borko Furht, Florida Atlantic University, Boca Raton, USA
V.S. Subrahmanian, University of Maryland, College Park, USA
Martial Hebert, Carnegie Mellon University, Pittsburgh, USA
Katsushi Ikeuchi, University of Tokyo, Tokyo, Japan
Bruno Siciliano, Università di Napoli Federico II, Napoli, Italy
Sushil Jajodia, George Mason University, Fairfax, USA
Newton Lee, Newton Lee Laboratories, LLC, Tujunga, USA

More information about this series at http://www.springer.com/series/10028

Mu Zhang • Heng Yin

Android Application Security

A Semantics and Context-Aware Approach

 Springer

Mu Zhang
Computer Security Department
NEC Laboratories America, Inc.
Princeton, NJ, USA

Heng Yin
University of California, Riverside
Riverside, CA, USA

ISSN 2191-5768   ISSN 2191-5776   (electronic)
SpringerBriefs in Computer Science
ISBN 978-3-319-47811-1   ISBN 978-3-319-47812-8   (eBook)
DOI 10.1007/978-3-319-47812-8

Library of Congress Control Number: 2016959410

Printed on acid-free paper

This Springer imprint is published by Springer Nature
The registered company is Springer International Publishing AG
The registered company address is: Gewerbestrasse 11, 6330 Cham, Switzerland

The authors would like to dedicate this book to their beloved families and friends and to those who overcome their frustration and persevere with resubmitting papers to top-tier computer security conferences.

Preface

This book is an introduction to the cutting-edge technologies for discovery, diagnosis, and defense of emerging security problems in modern Android applications.

With great power comes great threat. Recently, due to the popularity of Android smartphones, Android apps have attracted varieties of cyber attacks: some involve advanced anti-detection techniques; some exploit "genetic" defects in Android programs; some cover up identity theft with camouflage; some trick end users to fall into a trap using intriguing but misleading language. To defeat malicious attempts, researchers strike back. Many traditional techniques have been studied and practiced: malware classification, taint analysis, access control, etc. Yet, intrusive techniques also advance, and, unfortunately, existing defenses fall short, fundamentally due to the lack of sufficient interpretation of Android application behaviors.

To address this limitation, we look at the problem from a different angle. Android apps, no matter good, bad, or vulnerable, are in fact software programs. Their functionality is concretized through semantically meaningful code and varies under different circumstances. This reveals two essential factors for understanding Android application, semantics and contexts, which, we believe, are also the key to tackle security problems in Android apps. As a result, we have developed a series of semantics and context-aware techniques to fight against Android security threats. We have applied our idea to four significant areas, namely, malware detection, vulnerability patching, privacy leakage mitigation, and misleading app descriptions. This will be elaborated through the whole book.

Intended Audience

This book is suitable for security professionals and researchers. It will also be useful for graduate students who are interested in mobile application security.

Acknowledgments

The authors would like to thank Lok, Aravind, Andrew, Qian, Xunchao, Yue, Rundong, Jinghan, Manju, Eknath, and Curtis for the stimulating discussions and generous support.

Princeton, NJ, USA   Mu Zhang
Riverside, CA, USA   Heng Yin
September 2016

Contents

Chapter 1
Introduction

Abstract Along with the boom of Android apps come severe security challenges. Existing techniques fall short when facing emerging security problems in Android applications, such as zero-day or polymorphic malware, deep and complex vulnerabilities, privacy leaks and insecure app descriptions. To fight these threats, we have proposed a semantics and context aware approach, and designed and developed a series of advanced techniques.

1.1 Security Threats in Android Applications

Android has dominated the smartphone market and become the most popular operating system for mobile devices. In the meantime, security threats in Android apps have also quickly increased. In particular, four major classes of problems, *malware*, *program vulnerabilities*, *privacy leaks* and *insecure app descriptions*, bring considerable challenges to Android application security. Although a great deal of research efforts have been made to address these threats, they have fundamental limitations and thus cannot solve the problems.

1.1.1 Malware Attacks

Malicious apps have been increasingly exponentially according to McAfee's annual report [1]. Malware steals and pollutes sensitive information, executes attacker specified commands, or even totally roots and subverts the mobile devices. Unfortunately, existing automated Android malware detection and classification methods can be evaded both in theory and practice. Signature-based detections [2, 3] barely look for specific bytecode patterns, and therefore are easily evaded by bytecode-level transformation attacks [4]. In contrast, machine learning-based approaches [5–7] extract features from simple, external and isolated symptoms of an application (e.g., permission requests and individual API calls). The extracted features are thus associated with volatile application syntax, rather than high-level and robust program semantics. As a result, these detectors are also susceptible to evasion.

© The Author(s) 2016
M. Zhang, H. Yin, *Android Application Security*, SpringerBriefs in Computer
Science, DOI 10.1007/978-3-319-47812-8_1

1.1.2 Software Vulnerabilities

Apps may also contain security vulnerabilities, such as privilege escalation [8], capability leaks [9], permission re-delegation [10], component hijacking [11], content leaks & pollution [12] and inter-component communication vulnerabilities [13, 14]. These vulnerabilities are largely detected via automated static analysis [9, 11–14] to guarantee the scalability and satisfactory code coverage. However, static analysis is conservative in nature and may raise false positives. Therefore, once a "potential" vulnerability is discovered, it first needs to be confirmed; once it is confirmed, it then needs to be patched. Nevertheless, it is fairly challenging to programmatically accomplish these two tasks because it requires automated interpretation of program semantics. So far, upon receiving a discovered vulnerability, the developer has no choice but to manually confirm if the reported vulnerability is real. It may also be nontrivial for the (often inexperienced) developer to properly fix the vulnerability and release a patch for it. Thus, these discovered vulnerabilities may not be addressed for long time or not addressed at all, leaving a big time window for attackers to exploit these vulnerabilities.

1.1.3 Information Leakage

Information leakage is prevailing in both malware and benign applications. To address privacy leaks, prior efforts are made to perform both dynamic and static information flow analyses. Dynamic taint tracking based approaches, including DroidScope [15], VetDroid [16] and TaintDroid [17], can accurately detect information exfiltration at runtime, but incur significant performance overhead and suffer from insufficient code coverage. Conversely, static program analyses (e.g., FlowDroid [18] and AmanDroid [19]) can achieve high efficiency and coverage, but are only able to identify "potential" leakage and may cause considerable false positives. Besides, both approaches merely discover the presence of private data transmission, but do not consider how and why the transmission actually happens. Due to the lack of "context" information, these detectors cannot explain the cause of detected "privacy leaks". Thus, they cannot differentiate legitimate usage of private data (e.g., Geo-location required by Google Map) from true data leakage, and may yield severe false alarms.

1.1.4 Insecure Descriptions

Unlike traditional desktop systems, Android provides end users with an opportunity to proactively accept or deny the installation of any app to the system. As a result, it is essential that the users become aware of each app's behaviors so as to make appropriate decisions. To this end, Android markets directly present the consumers with two classes of information regarding each app: (1) permissions requested by the app and (2) textual descriptions provided by the developer. However, neither can serve the needs. Permissions are not only hard to understand [20] but also incapable of explaining how the requested permissions are used. For instance, both a benign navigation app and a spyware instance of the same app can require the same permission to access GPS location, yet use it for completely different purposes. While the benign app delivers GPS data to a legitimate map server upon the user's approval, the spyware instance can periodically and stealthily leak the user's location information to an attacker's site. Due to the lack of context clues, a user is not able to perceive such differences via the simple permission enumeration. Textual descriptions provided by developers are not security-centric. There exists very little incentive for app developers to describe their products from a security perspective, and it is still a difficult task for average developers to write dependable descriptions. Besides, malware authors deliberately deliver misleading descriptions so as to hide malice from innocent users. Previous studies [21, 22] have revealed that the existing descriptive texts are deviated considerably from requested permissions. As a result, developer-driven description generation cannot be considered trustworthy.

1.2   A Semantics and Context Aware Approach to Android Application Security

The fundamental problem of existing defense techniques lies in the fact that they are not aware of program semantics or contexts. To direct address the threats to Android application security, we propose a semantics and context-aware approach, which is more effective for malware detection and privacy protection, more usable to improve the security-awareness of end users and can address sophisticated software vulnerabilities that previously cannot be solved. Particularly, we propose four new techniques to address the specific security problems.

(1) **Android Malware Classification.** To battle malware polymorphism and zero-day malware, we extract contextual API dependency graphs of each Android app via program analysis, assign different weights to the nodes in the graphs using learning-based technique, measure the weighted graph similarity, and further use the similarity score to construct feature vectors.

(2) **Automatic Patch Generation for Component Hijacking Vulnerabilities.** To fix detected vulnerabilities, we perform static analysis to discover the small portion of code that leads to component hijacking attacks, and then selectively insert patch statements into the vulnerable program so as to defeat the exploitations at runtime.

(3) **Privacy Policy Enforcement.** To defeat privacy leakage without compromising legitimate functionality, we first rewrite the privacy breaching app to insert code that tracks sensitive information flow and enforces privacy polices, and then at

runtime associate the policies to specific contexts via modeling the user reaction to each incident of privacy violation.

(4) **Automated Generation of Security-Centric Descriptions.** To improve the security sensitivity of app descriptions, we first retrieve the behavior graphs from each Android app to reflect its security-related operations, then compress the graphs by collapsing the common patterns so as to produce more succinct ones, and finally translate the compressed graphs into concise and human-readable descriptions.

References

1. McAfee Labs Threats report Fourth Quarter (2013) http://www.mcafee.com/us/resources/reports/rp-quarterly-threat-q4-2013.pdf
2. Zhou Y, Wang Z, Zhou W, Jiang X (2012) Hey, you, get off of my market: detecting malicious apps in official and alternative android markets. In: Proceedings of 19th annual network and distributed system security symposium (NDSS)
3. Grace M, Zhou Y, Zhang Q, Zou S, Jiang X (2012) RiskRanker: scalable and accurate zero-day android malware detection. In: Proceedings of the 10th international conference on mobile systems, applications and services (MobiSys)
4. Rastogi V, Chen Y, Jiang X (2013) DroidChameleon: evaluating android anti-malware against transformation attacks. In: Proceedings of the 8th ACM symposium on InformAtion, computer and communications security (ASIACCS)
5. Peng H, Gates C, Sarma B, Li N, Qi Y, Potharaju R, Nita-Rotaru C, Molloy I (2012) Using probabilistic generative models for ranking risks of android apps. In: Proceedings of the 2012 ACM conference on computer and communications security (CCS)
6. Aafer Y, Du W, Yin H (2013) DroidAPIMiner: mining API-level features for robust malware detection in android. In: Proceedings of the 9th international conference on security and privacy in communication networks (SecureComm)
7. Arp D, Spreitzenbarth M, Hübner M, Gascon H, Rieck K (2014) Drebin: efficient and explainable detection of android malware in your pocket. In: Proceedings of the 21th annual network and distributed system security symposium (NDSS)
8. Davi L, Dmitrienko A, Sadeghi AR, Winandy M (2011) Privilege escalation attacks on android. In: Proceedings of the 13th international conference on Information security. Berlin/Heidelberg
9. Grace M, Zhou Y, Wang Z, Jiang X (2012) Systematic detection of capability leaks in stock android smartphones. In: Proceedings of the 19th network and distributed system security symposium
10. Felt AP, Wang HJ, Moshchuk A, Hanna S, Chin E (2011) Permission re-delegation: attacks and defenses. In: Proceedings of the 20th USENIX security symposium
11. Lu L, Li Z, Wu Z, Lee W, Jiang G (2012) CHEX: statically vetting android apps for component hijacking vulnerabilities. In: Proceedings of the 2012 ACM conference on computer and communications security (CCS)
12. Zhou Y, Jiang X (2013) Detecting passive content leaks and pollution in android applications. In: Proceedings of the 20th network and distributed system security symposium
13. Chin E, Felt AP, Greenwood K, Wagner D (2011) Analyzing inter-application communication in android. In: Proceedings of the 9th international conference on mobile systems, applications, and services (MobiSys)
14. Octeau D, McDaniel P, Jha S, Bartel A, Bodden E, Klein J, Traon YL (2013) Effective inter-component communication mapping in android with epicc: an essential step towards holistic security analysis. In: Proceedings of the 22nd USENIX security symposium

15. Yan LK, Yin H (2012) DroidScope: seamlessly reconstructing OS and Dalvik semantic views for dynamic android malware analysis. In: Proceedings of the 21st USENIX security symposium
16. Zhang Y, Yang M, Xu B, Yang Z, Gu G, Ning P, Wang XS, Zang B (2013) Vetting undesirable behaviors in android apps with permission use analysis. In: Proceedings of the 20th ACM conference on computer and communications security (CCS)
17. Enck W, Gilbert P, Chun BG, Cox LP, Jung J, McDaniel P, Sheth AN (2010) TaintDroid: an information-flow tracking system for realtime privacy monitoring on smartphones. In: Proceedings of the 9th USENIX symposium on operating systems design and implementation (OSDI)
18. Arzt S, Rasthofer S, Fritz C, Bodden E, Bartel A, Klein J, Traon YL, Octeau D, McDaniel P (2014) FlowDroid: precise context, flow, field, object-sensitive and lifecycle-aware taint analysis for android apps. In: Proceedings of the 35th ACM SIGPLAN conference on programming language design and implementation (PLDI)
19. Wei F, Roy S, Ou X, Robby (2014) Amandroid: a precise and general inter-component data flow analysis framework for security vetting of android apps. In: Proceedings of the 21th ACM conference on computer and communications security (CCS). Scottsdale
20. Felt AP, Ha E, Egelman S, Haney A, Chin E, Wagner D (2012) Android permissions: user attention, comprehension, and behavior. In: Proceedings of the eighth symposium on usable privacy and security (SOUPS)
21. Pandita R, Xiao X, Yang W, Enck W, Xie T (2013) WHYPER: towards automating risk assessment of mobile applications. In: Proceedings of the 22nd USENIX conference on security
22. Qu Z, Rastogi V, Zhang X, Chen Y, Zhu T, Chen Z (2014) Autocog: measuring the description-to-permission fidelity in android applications. In: Proceedings of the 21st conference on computer and communications security (CCS)

Chapter 2
Background

Abstract Android applications are developed on top of Android framework and therefore bear particular features compared to traditional desktop software. In the meantime, due to the unique design and implementation, Android apps are threatened by emerging cyber attacks that target at mobile operating systems. As a result, security researchers have made considerable efforts to discover, mitigate and defeat these threats.

2.1 Android Application

Android is a popular operating system for mobile devices. It dominates in the battle to be the top smartphone system in the world, and ranked as the top smartphone platform with 52 % market share (71.1 million subscribers) in Q1 2013. The success of Android is also reflected from the popularity of its applications. Tens of thousands of Android apps become available in Google Play while popular apps (e.g., Adobe Flash Player 11) have been downloaded and installed over 100 million times.

Android apps are developed using mostly Java programming language, with the support of Android Software Development Kit (SDK). The source code is first compiled into Java classes and further compiled into a Dalvik executable (i.e., DEX file) via dx tool. Then, the DEX program and other resource files (e.g., XML layout files, images) are assembled into the same package, called an APK file. The APK package is later submitted to the Android app markets (e.g., Google Play Store) with the developer's descriptions in text and other formats. An app market serves as the hub to distribute the application products, while consumers can browse the market and purchase the APK files. Once a APK file is downloaded and installed to a mobile device, the Dalvik executable will be running within a Dalvik virtual machine (DVM).

© The Author(s) 2016
M. Zhang, H. Yin, *Android Application Security*, SpringerBriefs in Computer
Science, DOI 10.1007/978-3-319-47812-8_2

2.1.1 *Android Framework API*

While an APK file is running in DVM, the Android framework code is also loaded and executed in the same domain. As a matter of fact, a Dalvik executable merely acts as a plug-in to the framework code, and a large portion of program execution happens within the Android framework.

A DEX file interacts with the Android framework via Application Programming Interface (API). These APIs are provided to the developers through Android SDK. From developers' perspective, Android API is the only channel for them to communicate with the underlying system and enable critical functionalities. Due to the nature of mobile operating system, Android offers a broad spectrum of APIs that are specific to smartphone capabilities. For instance, an Android app can programmatically send SMS messages via `sendTextMessage()` API or retrieve user's geographic location through `getLastKnownLocation()`.

2.1.2 *Android Permission*

Sensitive APIs are protected by Android permissions. Android exercises an install-time permission system. To enable the critical functionalities in an app, a developer has to specify the needs for corresponding permissions in a manifest file `AndroidManifest.xml`. Once an end user agrees to install the app, the required permissions are granted. At runtime, permission checks are enforced at both framework and system levels to ensure that an app has adequate privileges to make critical API calls.

There exist two major limitations for this permission system. Firstly, once certain permission is granted to an app at the install time, there is no easy way to revoke it at runtime. Secondly and more importantly, the permission enforcement is fundamentally a single-point check and thus lacking continuous protection. If an application can pass a checkpoint and retrieve sensitive data via a critical API call, it can use the data without any further restrictions.

2.1.3 *Android Component*

Android framework also provides a special set of APIs that are associated to Android components. Components in Android are the basic building units for apps. In particular, there exist four types of components in Android: `Activity`, `Service`, `Broadcast Receiver`, and `Content Provider`. An Activity class takes care of creating the graphical user interface (GUI) and directly interacts with the end user. A Service, in contrast, performs non-interactive longer-running operations in background while accepting service requests from other apps or app components. A Broadcast Receiver is component that listens to and processes system-wide broadcast messages. A Content Provider encapsulates data content and shares it with multiple components via a unified interface.

Components communicate with one another via Intents. An Intent with certain ACTION code, target component and payload data indicates a specific operation to be performed. For example, with different target parameter, an Intent can be used to launch an Activity, request a Service or send a message to any interested Broadcast Receiver. A developer can create custom permissions to protect components from receiving Intents from an arbitrary sender. However, such a simple mechanism cannot rule out malicious Intent communication because it does not prevent a malicious app author from requesting the same custom permission at install time.

2.1.4 Android App Description

Once an Android app has been developed, it is delivered to the app markets along with the developer's descriptions. Developers are usually interested in describing the app's functionalities, unique features, special offers, use of contact information, etc. Nevertheless, they are not motivated to explain the security risks behind the sensitive app functions. Prior studies [33, 36] have revealed significant inconsistencies between what the app is claimed to do and what the app actually does. This indicates that a majority of apps exercise undeclared sensitive functionalities beyond the users' expectation. Such a practice may not necessarily be malicious, but it does provide a potential window for attacks to exploit.

To mitigate this problem, Android markets also explains to users, in natural language, what permissions are required by an app. The goal is to help users understand the program behaviors so as to avoid security risks. However, such a simple explanation is still too technical for average users to comprehend. Besides, a permission list does not illustrate how permissions are used by an app. As an example, if an application first retrieves user's phone number and then sends it to a remote server, it in fact uses two permissions, READ_PHONE_STATE and INTERNET in a collaborative manner. Unfortunately, the permission list can merely inform that two independent permissions have been used.

2.2 Android Malware Detection

The number of new Android malware instances has grown exponentially in recent years. McAfee reports [28] that 2.47 million new mobile malware samples were collected in 2013, which represents a 197 % increase over 2012. Greater and greater amounts of manual effort are required to analyze the increasing number of new malware instances. This has led to a strong interest in developing methods to automate the malware analysis process.

Existing automated Android malware detection and classification methods fall into two general categories: (1) signature-based and (2) machine learning-based. Signature-based approaches [17, 54] look for specific patterns in the bytecode and API calls, but they are easily evaded by bytecode-level transformation attacks [37]. Machine learning-based approaches [1, 2, 34] extract features from an application's behavior (such as permission requests and critical API calls) and apply standard machine learning algorithms to perform binary classification. Because the extracted features are associated with application syntax, rather than program semantics, these detectors are also susceptible to evasion.

2.2.1 Signature Detection and Malware Analysis

Previous studies were focused on large-scale and light-weight detection of malicious or dangerous Android apps. DroidRanger [54] proposed permission-based footprinting and heuristics-based schemes to detect new samples of known malware families and identify certain behaviors of unknown malicious families, respectively. Risk-Ranker [17] developed an automated system to uncover dangerous app behaviors, such as root exploits, and assess potential security risks. Kirin [11] proposed a security service to certify apps based upon predefined security specifications. WHY-PER [33] leveraged Natural Language Processing and automated risk assessment of mobile apps by revealing discrepancies between application descriptions and their true functionalities. Efforts were also made to pursue in-depth analysis of malware and application behaviors. TaintDroid [12], DroidScope [47] and VetDroid [51] conducted dynamic taint analysis to detect suspicious behaviors during runtime. Ded [13], CHEX [25], PEG [6], and FlowDroid [3] exercised static dataflow analysis to identify dangerous code in Android apps. The effectiveness of these approaches depends upon the quality of human crafted detection patterns specific to certain dangerous or vulnerable behaviors.

2.2.2 Android Malware Classification

Many efforts have also been made to automatically classify Android malware via machine learning. Peng et al. [34] proposed a permission-based classification approach and introduced probabilistic generative models for ranking risks for Android apps. Juxtapp [18] performed feature hashing on the opcode sequence to detect malicious code reuse. DroidAPIMiner [1] extracted Android malware features at the API level and provided light-weight classifiers to defend against malware installations. DREBIN [2] took a hybrid approach and considered both

Android permissions and sensitive APIs as malware features. To this end, it performed broad static analysis to extract feature sets from both manifest files and bytecode programs. It further embedded all feature sets into a joint vector space. As a result, the features contributing to malware detection can be analyzed geometrically and used to explain the detection results. Despite the effectiveness and computational efficiency, these machine learning based approaches extract features from solely external symptoms and do not seek an accurate and complete interpretation of app behaviors.

2.3 Android Application Vulnerabilities

Although the permission-based sandboxing mechanism enforced in Android can effectively confine each app's behaviors by only allowing the ones granted with corresponding permissions, a vulnerable app with certain critical permissions can perform security-sensitive behaviors on behalf of a malicious app. It is so called confused deputy attack. This kind of security vulnerabilities can present in numerous forms, such as privilege escalation [8], capability leaks [16], permission re-delegation [14], content leaks and pollution [53], component hijacking [25], etc.

Prior work primarily focused on automatic discovery of these vulnerabilities. Once a vulnerability is discovered, it can be reported to the developer and a patch is expected. Some patches can be as simple as placing a permission validation at the entry point of an exposed interface (to defeat privilege escalation [8] and permission re-delegation [14] attacks), or withholding the public access to the internal data repositories (to defend against content leaks and pollution [53]), the fixes to the other problems may not be so straightforward.

2.3.1 Component Hijacking Vulnerabilities

In particular, component hijacking may fall into the latter category. When receiving a manipulated input from a malicious Android app, an app with a component hijacking vulnerability may exfiltrate sensitive information or tamper with the sensitive data in a critical data repository on behalf of the malicious app. In other words, a dangerous information flow may happen in either an outbound or inbound direction depending on certain external conditions and/or the internal program state.

A prior effort has been made to perform static analysis to discover *potential* component hijacking vulnerabilities [25]. Static analysis is known to be conservative in nature and may raise false positives. For instance, static analysis may find a viable execution path for information flow, which may never be reached in actual program execution; static analysis may find that interesting information is stored in some elements in a database, and thus has to conservatively treat the entire database as sensitive. As a result, upon receiving a discovered vulnerability, the developer has to manually confirm if the reported vulnerability is real. However, it is nontrivial for average developers to properly fix the vulnerability and release a patch.

2.3.2 Automatic Patch and Signature Generation

While an automated patching method is still lacking for vulnerable Android apps, a series of studies have been made to automatically generate patch for conventional client-server programs. AutoPaG [23] analyzes the program source code and identifies the root cause for out-of-bound exploit, and thus creates a fine-grained source code patch to temporarily fix it without any human intervention. IntPatch [50] utilizes classic type theory and dataflow analysis framework to identify potential integer-overflow-to-buffer-overflow vulnerabilities, and then instruments programs with runtime checks. Sidiroglou and Keromytis [39] rely on source code transformations to quickly apply automatically created patches to vulnerable segments of the targeted applications, that are subject to zero-day worms. Newsome et al. [31] propose an execution-based filter which filters out attacks on a specific vulnerability based on the vulnerable program's execution trace. ShieldGen [7] generates a data patch or a vulnerability signature for an unknown vulnerability, given a zero-day attack instance. Razmov and Simon [38] automate the filter generation process based on a simple formal description of a broad class of assumptions about the inputs to an application.

2.3.3 Bytecode Rewriting

In principle, these aforementioned patching techniques can be leveraged to address the vulnerabilities in Android apps. Nevertheless, to fix an Android app, a specific bytecode rewriting technique is needed to insert patch code into the vulnerable programs. Previous studies have utilized this technique to address varieties of problems. The Privacy Blocker application [35] performs static analysis of application binaries to identify and selectively replace requests for sensitive data with hard-coded shadow data. I-ARM-Droid [9] rewrites Dalvik bytecode to interpose on all the API invocations and enforce the desired security policies. Aurasium [46] repackages Android apps to sandbox important native APIs so as to monitor security and privacy violations. Livshits and Jung [24] implement a graph-theoretic algorithm to place mediation prompts into bytecode program and thus protect resource access. However, due the simplicity of the target problems, prior work did not attempt to rewrite the bytecode program in an extensive fashion. In contrast, to address sophisticated vulnerabilities, such as component hijacking, a new machinery has to be developed, so that inserted patch code can effectively monitor and control sensitive information flow in apps.

2.3.4 Instrumentation Code Optimization

The size of a rewritten program usually increases significantly. Thus, an optimization phase is needed. Several prior studies attempted to reduce code instrumentation overhead by performing various static analysis and optimizations. To find error patterns in Java source code, Martin et al. optimized dynamic instrumentation by performing static pointer alias analysis [27]. To detect numerous software attacks, Xu et al. inserted runtime checks to enforce various security policies in C source code, and remove redundant checks via compiler optimizations [45]. As a comparison, due to the limited resources on mobile devices, there exists an even more strict restriction for app size. Therefore, a novel method is necessary to address this new challenge.

2.4 Privacy Leakage in Android Apps

While powerful Android APIs facilitate versatile functionalities, they also arouse privacy concerns. Previous studies [12, 13, 19, 44, 52, 54] have exposed that both benign and malicious apps are stealthily leaking users' private information to remote servers. Efforts have also been made to detect and analyze privacy leakage either statically or dynamically [12, 13, 15, 22, 25, 26, 48]. Nevertheless, a good solution to defeat privacy leakage at runtime is still lacking.

2.4.1 Privacy Leakage Detection

Egele et al. [10] studied the privacy threats in iOS applications. They proposed PiOS, a static analysis tool to detect privacy leaks in Mach-O binaries. TaintDroid is a dynamic analysis tool for detecting and analyzing privacy leaks in Android applications [12]. It modifies Dalvik virtual machine and dynamically instruments Dalvik bytecode instructions to perform dynamic taint analysis. Enck et al. [13] proposed a static analysis approach to study privacy leakage as well. They convert a Dalvik executable to Java source code and leverage a commercial Java source code analysis tool Fortify360 [20] to detect surreptitious data flows. CHEX [25] is designed to vet Android apps for component hijacking vulnerabilities and is essentially capable of detecting privacy leakage. It converted Dalvik bytecode to WALA [43] SSA IR, and conducted static dataflow analysis with WALA framework. AndroidLeaks [15] is a static analysis framework, which also leverages WALA, and identifies potential leaks of personal information in Android applications on a large scale. Mann et al. [26] analyzed the Android API for possible sources and sinks of private data and thus identified exemplary privacy policies. All the existing detection methods fundamental cause significant false alarms because they cannot differentiate legitimate use of sensitive data from authentic privacy leakage. Though effective in terms of privacy protection, these approaches did not attempt to preserve the system usability.

2.4.2 Privacy Leak Mitigation

Based on TaintDroid, Hornyack et al. [19] proposed AppFence to further mitigate privacy leaks. When TaintDroid discovers the data dependency between source and sink, AppFence enforces privacy policies, either at source or sink, to protect sensitive information. At source, it may provide the app with fake information instead of the real one; at sink, it can block sending APIs. To take usability into consideration, the authors proposed multiple access control rules and conducted empirical studies to find the optimal policies in practice.

The major limitation of AppFence is the lack of efficiency. AppFence requires modifications in the Dalvik virtual machine to track information flow and incurs considerable performance overhead (14 % on average according to TaintDroid [12]). Besides, the deployment is also challenging. For one thing, end users have to re-install the operating system on their mobile device to enable AppFence. For another, once the Android OS upgrades to a new version, AppFence needs to be re-engineered to work with the novel mechanisms.

2.4.3 Information Flow Control

Though AppFence is limited by its efficiency and deployment, it demonstrates that it is feasible to leverage Information-Flow Control (IFC) technique to address the privacy leakage problem in Android apps. In fact, IFC has been studied on different contexts. Chandra and Franz [5] implement an information flow framework for Java virtual machine which combines static analysis to capture implicit flows. JFlow [30] extends the Java language and adds statically-checked information flow annotations. Jia et al. [21] proposes a component-level runtime enforcement system for Android apps. duPro [32] is an efficient user-space information flow control framework, which adopts software-based fault isolation to isolate protection domains within the same process. Zeng et al. [49] introduces an IRM-implementation framework at a compiler intermediate-representation (IR) level.

2.5 Text Analytics for Android Security

Recently, efforts have been made to study the security implications of textual descriptions for Android apps. WHYPER [33] used natural language processing technique to identify the descriptive sentences that are associated to permissions requests. It implemented a semantic engine to connect textual elements to Android permissions. AutoCog [36] further applied machine learning technique to automatically correlate the descriptive scripts to permissions, and therefore was able to assess description-to-permission fidelity of applications. These studies demonstrates the urgent need to bridge the gap between the textual description and security-related program semantics.

2.5.1 Automated Generation of Software Description

There exists a series of studies on software description generation for traditional Java programs. Sridhara et al. [40] automatically summarized method syntax and function logic using natural language. Later, they [41] improved the method summaries by also describing the specific roles of method parameters and integrating parameter descriptions. They presented heuristics to generate comments and describe the specific roles of different method parameters. Further, they [42] automatically identified high-level abstractions of actions in code and described them in natural language and attempted to automatically identify code fragments that implement high level abstractions of actions and express them as a natural language description. In the meantime, Buse [4] leveraged symbolic execution and code summarization technique to document program differences, and thus synthesize succinct human-readable documentation for arbitrary program differences. Moreno et al. [29] proposed a summarization tool which determines class and method stereotypes and uses them, in conjunction with heuristics, to select the information to be included in the class summaries. The goal of these studies is to improve the program comprehension for developers. As a result, they focus on documenting intra-procedural program logic and low-level code structures. On the contrary, they did not aim at depicting high-level program semantics and therefore cannot help end users to understand the risk of Android apps.

References

1. Aafer Y, Du W, Yin H (2013) DroidAPIMiner: mining API-level features for robust malware detection in android. In: Proceedings of the 9th international conference on security and privacy in communication networks (SecureComm)
2. Arp D, Spreitzenbarth M, Hübner M, Gascon H, Rieck K (2014) Drebin: efficient and explainable detection of android malware in your pocket. In: Proceedings of the 21th annual network and distributed system security symposium (NDSS)
3. Arzt S, Rasthofer S, Fritz C, Bodden E, Bartel A, Klein J, Traon YL, Octeau D, McDaniel P (2014) FlowDroid: precise context, flow, field, object-sensitive and lifecycle-aware taint analysis for android apps. In: Proceedings of the 35th ACM SIGPLAN conference on programming language design and implementation (PLDI)
4. Buse RP, Weimer WR (2010) Automatically documenting program changes. In: Proceedings of the IEEE/ACM international conference on automated software engineering (ASE)

5. Chandra D, Franz M (2007) Fine-grained information flow analysis and enforcement in a java virtual machine. In: Proceedings of the 23rd annual computer security applications conference (ACSAC)

6. Chen KZ, Johnson N, D'Silva V, Dai S, MacNamara K, Magrino T, Wu EX, Rinard M, Song D (2013) Contextual policy enforcement in android applications with permission event graphs. In: Proceedings of the 20th annual network and distributed system security symposium (NDSS)

7. Cui W, Peinado M, Wang HJ (2007) Shieldgen: automatic data patch generation for unknown vulnerabilities with informed probing. In: Proceedings of 2007 IEEE symposium on security and privacy

8. Davi L, Dmitrienko A, Sadeghi AR, Winandy M (2011) Privilege escalation attacks on android. In: Proceedings of the 13th international conference on Information security. Berlin/Heidelberg

9. Davis B, Sanders B, Khodaverdian A, Chen H (2012) I-ARM-Droid: a rewriting framework for in-app reference monitors for android applications. In: Proceedings of the mobile security technologies workshop

10. Egele M, Kruegel C, Kirda E, Vigna G (2011) PiOS: detecting privacy leaks in iOS applications. In: Proceedings of NDSS

11. Enck W, Ongtang M, McDaniel P (2009) On lightweight mobile phone application certification. In: Proceedings of the 16th ACM conference on computer and communications security (CCS)

12. Enck W, Gilbert P, Chun BG, Cox LP, Jung J, McDaniel P, Sheth AN (2010) TaintDroid: an information-flow tracking system for realtime privacy monitoring on smartphones. In: Proceedings of the 9th USENIX symposium on operating systems design and implementation (OSDI)

13. Enck W, Octeau D, McDaniel P, Chaudhuri S (2011) A study of android application security. In: Proceedings of the 20th USENIX Security Symposium

14. Felt AP, Wang HJ, Moshchuk A, Hanna S, Chin E (2011) Permission re-delegation: attacks and defenses. In: Proceedings of the 20th USENIX security symposium

15. Gibler C, Crussell J, Erickson J, Chen H (2012) AndroidLeaks: automatically detecting potential privacy leaks in android applications on a large scale. In: Proceedings of the 5th international conference on trust and trustworthy computing

16. Grace M, Zhou Y, Wang Z, Jiang X (2012) Systematic detection of capability leaks in stock android smartphones. In: Proceedings of the 19th network and distributed system security symposium

17. Grace M, Zhou Y, Zhang Q, Zou S, Jiang X (2012) RiskRanker: scalable and accurate zero-day android malware detection. In: Proceedings of the 10th international conference on mobile systems, applications and services (MobiSys)

18. Hanna S, Huang L, Wu E, Li S, Chen C, Song D (2012) Juxtapp: a scalable system for detecting code reuse among android applications. In: Proceedings of the 9th international conference on detection of intrusions and malware, and vulnerability assessment (DIMVA)

19. Hornyack P, Han S, Jung J, Schechter S, Wetherall D (2011) These aren't the droids you're looking for: retrofitting android to protect data from imperious applications. In: Proceedings of CCS

20. HP Fortify Source Code Analyzer (2016) http://www8.hp.com/us/en/software-solutions/static-code-analysis-sast/

21. Jia L, Aljuraidan J, Fragkaki E, Bauer L, Stroucken M, Fukushima K, Kiyomoto S, Miyake Y (2013) Run-time enforcement of information-flow properties on android (extended abstract). In: Computer Security–ESORICS 2013: 18th European symposium on research in computer security

22. Kim J, Yoon Y, Yi K, Shin J (2012) Scandal: static analyzer for detecting privacy leaks in android applications. In: Mobile security technologies (MoST)

23. Lin Z, Jiang X, Xu D, Mao B, Xie L (2007) AutoPAG: towards automated software patch generation with source code root cause identification and repair. In: Proceedings of the 2nd ACM symposium on information, computer and communications security

24. Livshits B, Jung J (2013) Automatic mediation of privacy-sensitive resource access in smartphone applications. In: Proceedings of the 22th USENIX security symposium
25. Lu L, Li Z, Wu Z, Lee W, Jiang G (2012) CHEX: statically vetting android apps for component hijacking vulnerabilities. In: Proceedings of the 2012 ACM conference on computer and communications security (CCS)
26. Mann C, Starostin A (2012) A framework for static detection of privacy leaks in android applications. In: Proceedings of the 27th annual ACM symposium on applied computing
27. Martin M, Livshits B, Lam MS (2005) Finding application errors and security flaws using PQL: a program query language. In: Proceedings of the 20th annual ACM SIGPLAN conference on object-oriented programming, systems, languages, and applications
28. McAfee Labs Threats report Fourth Quarter (2013) http://www.mcafee.com/us/resources/reports/rp-quarterly-threat-q4-2013.pdf
29. Moreno L, Aponte J, Sridhara G, Marcus A, Pollock L, Vijay-Shanker K (2013) Automatic generation of natural language summaries for java classes. In: Proceedings of the 2013 IEEE 21th international conference on program comprehension (ICPC)
30. Myers AC (1999) JFlow: practical mostly-static information flow control. In: Proceedings of the 26th ACM symposium on principles of programming languages (POPL)
31. Newsome J (2006) Vulnerability-specific execution filtering for exploit prevention on commodity software. In: Proceedings of the 13th symposium on network and distributed system security (NDSS)
32. Niu B, Tan G (2013) Efficient user-space information flow control. In: Proceedings of the 8th ACM symposium on information, computer and communications security
33. Pandita R, Xiao X, Yang W, Enck W, Xie T (2013) WHYPER: towards automating risk assessment of mobile applications. In: Proceedings of the 22nd USENIX conference on security
34. Peng H, Gates C, Sarma B, Li N, Qi Y, Potharaju R, Nita-Rotaru C, Molloy I (2012) Using probabilistic generative models for ranking risks of android apps. In: Proceedings of the 2012 ACM conference on computer and communications security (CCS)
35. Privacy Blocker (2016) http://privacytools.xeudoxus.com/
36. Qu Z, Rastogi V, Zhang X, Chen Y, Zhu T, Chen Z (2014) Autocog: measuring the description-to-permission fidelity in android applications. In: Proceedings of the 21st conference on computer and communications security (CCS)
37. Rastogi V, Chen Y, Jiang X (2013) DroidChameleon: evaluating android anti-malware against transformation attacks. In: Proceedings of the 8th ACM symposium on information, computer and communications security (ASIACCS)
38. Razmov V, Simon D (2001) Practical automated filter generation to explicitly enforce implicit input assumptions. In: Proceedings of the 17th annual computer security applications conference
39. Sidiroglou S and Keromytis AD (2005) Countering network worms through automatic patch generation. IEEE Secur Priv 3:41–49
40. Sridhara G, Hill E, Muppaneni D, Pollock L, Vijay-Shanker K (2010) Towards automatically generating summary comments for java methods. In: Proceedings of the IEEE/ACM international conference on automated software engineering (ASE)
41. Sridhara G, Pollock L, Vijay-Shanker K (2011) Generating parameter comments and integrating with method summaries. In: Proceedings of the 2011 IEEE 19th international conference on program comprehension (ICPC)
42. Sridhara G, Pollock L, Vijay-Shanker K (2011) Automatically detecting and describing high level actions within methods. In: Proceedings of the 33rd international conference on software engineering (ICSE)
43. T.J. Watson Libraries for Analysis (2015) http://wala.sourceforge.net/wiki/index.php/Main_Page
44. Wu C, Zhou Y, Patel K, Liang Z, Jiang X (2014) AirBag: boosting smartphone resistance to malware infection. In: Proceedings of the 21th annual network and distributed system security symposium (NDSS)

45. Xu W, Bhatkar S, Sekar R (2006) Taint-enhanced policy enforcement: a practical approach to defeat a wide range of attacks. In: Proceedings of the 15th conference on USENIX security symposium
46. Xu R, Sadi H, Anderson R (2012) Aurasium: practical policy enforcement for android applications. In: Proceedings of the 21th USENIX security symposium
47. Yan LK, Yin H (2012) DroidScope: seamlessly reconstructing OS and Dalvik semantic views for dynamic android malware analysis. In: Proceedings of the 21st USENIX security symposium
48. Yang Z, Yang M, Zhang Y, Gu G, Ning P, Wang XS (2013) AppIntent: analyzing sensitive data transmission in android for privacy leakage detection. In: Proceedings of the 20th ACM conference on computer and communications security (CCS)
49. Zeng B, Tan G, Erlingsson U (2013) Strato: a retargetable framework for low-level inlined-reference monitors. In: Proceedings of the 22th USENIX security symposium
50. Zhang C, Wang T, Wei T, Chen Y, Zou W (2010) IntPatch: automatically fix integer-overflow-to-buffer-overflow vulnerability at compile-time. In: Proceedings of the 15th European conference on research in computer security
51. Zhang Y, Yang M, Xu B, Yang Z, Gu G, Ning P, Wang XS, Zang B (2013) Vetting undesirable behaviors in android apps with permission use analysis. In: Proceedings of the 20th ACM conference on computer and communications security (CCS)
52. Zhou Y, Jiang X (2012) Dissecting android malware: characterization and evolution. In: Proceedings of the 33rd IEEE symposium on security and privacy. Oakland
53. Zhou Y, Jiang X (2013) Detecting passive content leaks and pollution in android applications. In: Proceedings of the 20th network and distributed system security symposium
54. Zhou Y, Wang Z, Zhou W, Jiang X (2012) Hey, you, get off of my market: detecting malicious apps in official and alternative android markets. In: Proceedings of 19th annual network and distributed system security symposium (NDSS)

Chapter 3
Semantics-Aware Android Malware Classification

Abstract The drastic increase of Android malware has led to a strong interest in developing methods to automate the malware analysis process. Existing automated Android malware detection and classification methods fall into two general categories: (1) signature-based and (2) machine learning-based. Signature-based approaches can be easily evaded by bytecode-level transformation attacks. Prior learning-based works extract features from application syntax, rather than program semantics, and are also subject to evasion. In this paper, we propose a novel semantic-based approach that classifies Android malware via dependency graphs. To battle transformation attacks, we extract a *weighted contextual API dependency graph* as program semantics to construct feature sets. To fight against malware variants and zero-day malware, we introduce graph similarity metrics to uncover homogeneous application behaviors while tolerating minor implementation differences. We implement a prototype system, *DroidSIFT*, in 23 thousand lines of Java code. We evaluate our system using 2200 malware samples and 13,500 benign samples. Experiments show that our signature detection can correctly label 93 % of malware instances; our anomaly detector is capable of detecting zero-day malware with a low false negative rate (2 %) and an acceptable false positive rate (5.15 %) for a vetting purpose.

3.1 Introduction

The number of new Android malware instances has grown exponentially in recent years. McAfee reports [1] that 2.47 million new mobile malware samples were collected in 2013, which represents a 197 % increase over 2012. Greater and greater amounts of manual effort are required to analyze the increasing number of new malware instances. This has led to a strong interest in developing methods to automate the malware analysis process.

Existing automated Android malware detection and classification methods fall into two general categories: (1) signature-based and (2) machine learning-based. Signature-based approaches [2, 3] look for specific patterns in the bytecode and API calls, but they are easily evaded by bytecode-level transformation attacks [4]. Machine learning-based approaches [5–7] extract features from an application's behavior (such as permission requests and critical API calls) and apply standard

© The Author(s) 2016   19
M. Zhang, H. Yin, *Android Application Security*, SpringerBriefs in Computer
Science, DOI 10.1007/978-3-319-47812-8_3

machine learning algorithms to perform binary classification. Because the extracted features are associated with application syntax, rather than program semantics, these detectors are also susceptible to evasion.

To directly address malware that evades automated detection, prior works distill program semantics into graph representations, such as control-flow graphs [8], data dependency graphs [9, 10] and permission event graphs [11]. These graphs are checked against manually-crafted specifications to detect malware. However, these detectors tend to seek an exact match for a given specification and therefore can potentially be evaded by polymorphic variants. Furthermore, the specifications used for detection are produced from known malware families and cannot be used to battle zero-day malware.

In this chapter, we propose a novel semantic-based approach that classifies Android malware via dependency graphs. To battle transformation attacks [4], we extract a weighted contextual API dependency graph as program semantics to construct feature sets. The subsequent classification then depends on more robust semantic-level behavior rather than program syntax. It is much harder for an adversary to use an elaborate bytecode-level transformation to evade such a training system. To fight against malware variants and zero-day malware, we introduce graph similarity metrics to uncover homogeneous essential application behaviors while tolerating minor implementation differences. Consequently, new or polymorphic malware that has a unique implementation, but performs common malicious behaviors, cannot evade detection.

To our knowledge, when compared to traditional semantics-aware approaches for desktop malware detection, we are the first to examine program semantics within the context of Android malware classification. We also take a step further to defeat malware variants and zero-day malware by comparing the similarity of these programs to that of known malware at the behavioral level.

We build a database of behavior graphs for a collection of Android apps. Each graph models the API usage scenario and program semantics of the app that it represents. Given a new app, a query is made for the app's behavior graphs to search for the most similar counterpart in the database. The query result is a similarity score which sets the corresponding element in the feature vector of the app. Every element of this feature vector is associated with an individual graph in the database.

We build graph databases for two sets of behaviors: benign and malicious. Feature vectors extracted from these two sets are then used to train two separate classifiers for anomaly detection and signature detection. The former is capable of discovering zero-day malware, and the latter is used to identify malware variants.

We implement a prototype system, *DroidSIFT*, in 23 thousand lines of Java code. Our dependency graph generation is built on top of Soot [12], while our graph similarity query leverages a graph matching toolkit[13] to compute graph edit distance. We evaluate our system using 2200 malware samples and 13,500 benign samples. Experiments show that our signature detection can correctly label 93 % malware instances; our anomaly detector is capable of detecting zero-day malware with a low false negative rate (2 %) and an acceptable false positive rate (5.15 %) for vetting purpose.

3.2 Overview

3.2.1 Problem Statement

An effective vetting process for discovering malicious software is essential for maintaining a healthy ecosystem in the Android app markets. Unfortunately, existing vetting processes are still fairly rudimentary. As an example, consider the Bouncer [14] vetting system that is used by the official Google Play Android market. Though the technical details of Bouncer are not publicly available, experiments by Oberheide and Miller [15] show that Bouncer performs dynamic analysis to examine apps within an emulated Android environment for a limited period of time. This method of analysis can be easily evaded by apps that perform emulator detection, contain hidden behaviors that are timer-based, or otherwise avoid triggering malicious behavior during the short time period when the app is being vetted. Signature detection techniques adopted by the current anti-malware products have also been shown to be trivially evaded by simple bytecode-level transformations [4].

We propose a new technique, *DroidSIFT*, illustrated in Fig. 3.1, that addresses these shortcomings and can be deployed as a replacement for existing vetting techniques currently in use by the Android app markets. This technique is based on static analysis, which is immune to emulation detection and is capable of analyzing the entirety of an app's code. Further, to defeat bytecode-level transformations, our static analysis is semantics-aware and extracts program behaviors at the semantic level.

Consequently, we are able to conduct two kinds of classifications: anomaly detection and signature detection. Upon receiving a new app submission, our vetting process will conduct anomaly detection to determine whether it contains behaviors that significantly deviate from the benign apps within our database. If such a deviation is discovered, a potential malware instance is identified. Then, we conduct further signature detection on it to determine if this app falls into any malware family within our signature database. If so, the app is flagged as malicious and bounced back to the developer immediately.

If the app passes this hurdle, it is still possible that a new malware species has been found. We bounce the app back to the developer with a detailed report when suspicious behaviors that deviate from benign behaviors are discovered,

Fig. 3.1 Deployment of DroidSIFT

and we request justification for the deviation. The app is approved only after the developer makes a convincing justification for the deviation. Otherwise, after further investigation, we may confirm it to indeed be a new malware species. By placing this information into our malware database, we further improve our signature detection to detect this new malware species in the future.

It is also possible to deploy our technique via a more ad-hoc scheme. For example, our detection mechanism can be deployed as a public service that allows a cautious app user to examine an app prior to its installation. An enterprise that maintains its own private app repository could utilize such a security service. The enterprise service conducts vetting prior to adding an app to the internal app pool, thereby protecting employees from apps that contain malware behaviors.

3.2.2 Architecture Overview

Figure 3.2 depicts the workflow of our graph-based Android malware classification. This takes the following steps:

(1) **Behavior Graph Generation.** Our malware classification considers graph similarity as the feature vector. To this end, we first perform static program analysis to transform Android bytecode programs into their corresponding graph representations. Our program analysis includes entry point discovery and call graph analysis to better understand the API calling contexts, and it leverages both forward and backward dataflow analysis to explore API dependencies and uncover any constant parameters. The result of this analysis is expressed via *Weighted Contextual API Dependency Graphs* that expose security-related behaviors of Android apps.

(2) **Scalable Graph Similarity Query.** Having generated graphs for both benign and malicious apps, we then query the graph database for the one graph most similar to a given graph. To address the scalability challenge, we utilize a bucket-based indexing scheme to improve search efficiency. Each bucket contains graphs bearing APIs from the same Android packages, and it is indexed

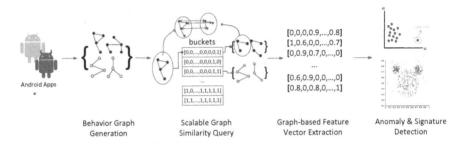

Fig. 3.2 Overview of DroidSIFT

with a bitvector that indicates the presence of such packages. Given a graph query, we can quickly seek to the corresponding bucket index by matching the package's vector to the bucket's bitvector. Once a matching bucket is located, we further iterate this bucket to find the best-matching graph. Finding the best-matching graph, instead of an exact match, is necessary to identify polymorphic malware.

(3) **Graph-based Feature Vector Extraction.** Given an app, we attempt to find the best match for each of its graphs from the database. This produces a similarity feature vector. Each element of the vector is associated with an existing graph in the database. This vector bears a non-zero similarity score in one element only if the corresponding graph is the best match to one of the graphs for the given app.

(4) **Anomaly and Signature Detection.** We have implemented a signature classifier and an anomaly detector. We have produced feature vectors for malicious apps, and these vectors are used to train the classifier for signature detection. The anomaly detection discovers zero-day Android malware, and the signature detector uncovers the type (family) of the malware.

3.3 Weighted Contextual API Dependency Graph

3.3.1 Key Behavioral Aspects

We consider the following aspects as essential when describing the semantic-level behaviors of an Android malware sample:

(1) **API Dependency.** API calls (including reflective calls to the private framework functions) indicate how an app interacts with the Android framework. It is essential to capture what API calls an app makes and the dependencies among those calls. Prior works on semantic- and behavior-based malware detection and classification for desktop environments all make use of API dependency information [9, 10]. Android malware shares the same characteristics.

(2) **Context.** An entry point of an API call is a program entry point that directly or indirectly triggers the call. From a user-awareness point of view, there are two kinds of entry points: user interfaces and background callbacks. Malware authors commonly exploit background callbacks to enable malicious functionalities without the user's knowledge. From a security analyst's perspective, it is a suspicious behavior when a typical user interactive API (e.g., AudioRecord.startRecording()) is called stealthily [11]. As a result, we must pay special attention to APIs activated from background callbacks.

(3) **Constant.** Constants convey semantic information by revealing the values of critical parameters and uncovering fine-grained API semantics. For instance, Runtime.exec() may execute varied shell commands, such as ps or chmod, depending upon the input string constant. Constant analysis

also discloses the data dependencies of certain security-sensitive APIs whose benign-ness is dependent upon whether an input is constant. For example, a sendTextMessage() call taking a constant premium-rate phone number as a parameter is a more suspicious behavior than the call to the same API receiving the phone number from user input via getText(). Consequently, it is crucial to extract information about the usage of constants for security analysis.

Once we look at app behaviors using these three perspectives, we perform similarity checking, rather than seeking an exact match, on the behavioral graphs. Since each individual API node plays a distinctive role in an app, it contributes differently to the graph similarity. With regards to malware detection, we emphasize security-sensitive APIs combined with critical contexts or constant parameters. We assign weights to different API nodes, giving greater weights to the nodes containing critical calls, to improve the "quality" of behavior graphs when measuring similarity. Moreover, the weight generation is automated. Thus, similar graphs have higher similarity scores by design.

3.3.2   Formal Definition

To address all of the aforementioned factors, we describe app behaviors using *Weighted Contextual API Dependency Graphs* (WC-ADG). At a high level, a WC-ADG consists of API operations where some of the operations have data dependencies. A formal definition is presented as follows.

Definition 1. A *Weighted Contextual API Dependency Graph* is a directed graph $G = (V, E, \alpha, \beta)$ over a set of API operations Σ and a weight space W, where:

- The set of vertices V corresponds to the contextual API operations in Σ;
- The set of edges $E \subseteq V \times V$ corresponds to the *data dependencies* between operations;
- The labeling function $\alpha : V \rightarrow \Sigma$ associates nodes with the labels of corresponding contextual API operations, where each label is comprised of 3 elements: API prototype, entry point and constant parameter;
- The labeling function $\beta : V \rightarrow W$ associates nodes with their corresponding weights, where $\forall w \in W$, $w \in R$, and R represents the space of real numbers.

3.3.3   A Real Example

Zitmo is a class of banking trojan malware that steals a user's SMS messages to discover banking information (e.g., mTANs). Figure 3.3 presents an example WC-ADG that depicts the malicious behavior of a Zitmo malware sample in a concise, yet complete, manner. This graph contains five API call nodes. Each node contains

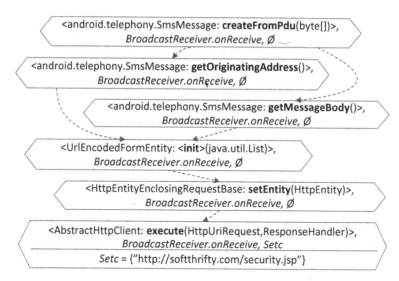

Fig. 3.3 WC-ADG of zitmo

the call's prototype, a set of any constant parameters, and the entry points of the call. Dashed arrows that connect a pair of nodes indicates that a data dependency exists between the two calls in those nodes.

By combining the knowledge of API prototypes with the data dependency information shown in the graph, we know that the app is forwarding an incoming SMS to the network. Once an SMS is received by the mobile phone, Zitmo creates an SMS object from the raw Protocol Data Unit by calling createFromPdu(byte[]). It extracts the sender's phone number and message content by calling getOriginating- Address() and getMessageBody(). Both strings are encoded into an UrlEncoded- FormEntity object and enclosed into HttpEntityEnclosingRequestBase by using the setEntity() call. Finally, this HTTP request is sent to the network via AbstractHttpClient.execute().

Zitmo variants may also exploit various other communication-related API calls for the sending purpose. Another Zitmo instance uses SmsManager.sendTextMess-age() to deliver the stolen information as a text message to the attacker's phone. Such variations motivate us to consider graph similarity metrics, rather than an exact matching of API call behavior, when determining whether a sample app is benign or malicious.

The context provided by the entry points of these API calls informs us that the user is not aware of this SMS forwarding behavior. These consecutive API invocations start within the entry point method onReceive() with a call to createFrom- Pdu(byte[]). onReceive() is a broadcast receiver registered by the app to receive incoming SMS messages in the background. Therefore, the createFromPdu(byte[]) and subsequent API calls are activated from a non-user-interactive entry point and are hidden from the user.

Constant analysis of the graph further indicates that the forwarding destination is suspicious. The parameter of `execute()` is neither the sender (i.e., the bank) nor any familiar parties from the contacts. It is a constant URL belonging to an unknown third-party.

3.3.4 Graph Generation

We have implemented a graph generation tool on top of Soot [12] in 20 k lines of code. This tool examines an Android app to conduct entry point discovery and perform context-sensitive, flow-sensitive, and interprocedural dataflow analyses. These analyses locate API call parameters and return values of interest, extract constant parameters, and determine the data dependencies among the API calls.

3.3.4.1 Entry Point Discovery

Entry point discovery is essential to revealing whether the user is aware that a certain API call has been made. However, this identification is not straightforward. Consider the callgraph seen in Fig. 3.4. This graph describes a code snippet that registers a `onClick()` event handler for a button. From within the event handler, the code starts a thread instance by calling `Thread.start()`, which invokes the `run()` method implementing `Runnable.run()`. The `run()` method passes an `android.os.Message` object to the message queue of the hosting thread via `Handler.sendMessage()`. A `Handler` object created in the same thread is then bound to this message queue and its `Handler.handleMessage()` callback will process the message and later execute `sendTextMessage()`.

The sole entry point to the graph is the user-interactive callback `onClick()`. However, prior work [16] on the identification of program entry points does not consider asynchronous calls and recognizes all three callbacks in the program as individual entry points. This confuses the determination of whether the user is aware

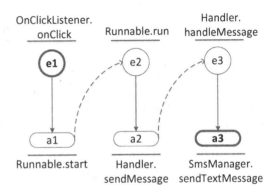

Fig. 3.4 Callgraph for asynchronously sending an SMS message. "e" and "a" stand for "event handler" and "action" respectively

---

Algorithm 1 Entry Point Reduction for Asynchronous Callbacks

M_{entry} ← {Possible entry point callback methods}
CM_{async} ← {Pairs of ($BaseClass$, $RunMethod$) for asynchronous calls in framework}
RS_{async} ← {Map from $RunMethod$ to $StartMethod$ for asynchronous calls in framework}
for $m_{entry} \in M_{entry}$ **do**
    c ← the class declaring m_{entry}
    $base$ ← the base class of c
    if ($base$, m_{entry}) $\in CM_{async}$ **then**
        m_{start} ← Lookup(m_{entry}) in RS_{async}
        for $\forall$ $call$ to m_{start} **do**
            r ← "this" reference of $call$
            $PointsToSet$ ← PointsToAnalysis(r)
            if $c \in PointsToSet$ **then**
                $M_{entry} = M_{entry} - \{m_{entry}\}$
                BuildDependencyStub(m_{start}, m_{entry})
            end if
        end for
    end if
end for
output M_{entry} as reduced entry point set

---

that an API call has been made in response to a user-interactive callback. To address this limitation, we propose Algorithm 1 to remove any potential entry points that are actually part of an asynchronous call chain with only a single entry point.

Algorithm 1 accepts three inputs and provides one output. The first input is M_{entry}, which is a set of possible entry points. The second is CM_{async}, which is a set of ($BaseClass$, $RunMethod$) pairs. $BaseClass$ represents a top-level asynchronous base class (e.g., Runnable) in the Android framework and $RunMethod$ is the asynchronous call target (e.g., Runnable.run()) declared in this class. The third input is RS_{async}, which maps $RunMethod$ to $StartMethod$. $RunMethod$ and $StartMethod$ are the callee and caller in an asynchronous call (e.g., Runnable.run() and Runnable.start()). The output is a reduced M_{entry} set.

We compute the M_{entry} input by applying the algorithm proposed by Lu et al. [16], which discovers all reachable callback methods defined by the app that are intended to be called only by the Android framework. To further consider the logical order between Intent senders and receivers, we leverage Epicc [17] to resolve the inter-component communications and then remove the Intent receivers from M_{entry}.

Through examination of the Android framework code, we generate a list of 3-tuples consisting of $BaseClass$, $RunMethod$ and $StartMethod$. For example, we capture the Android-specific calling convention of AsyncTask with AsyncTask.on- PreExecute() being triggered by AsyncTask.execute(). When a new asynchronous call is introduced into the framework code, this list is updated to include the change. Table 3.1 presents our current model for the calling convention of top-level base asynchronous classes in Android framework.

Given these inputs, our algorithm iterates over M_{entry}. For every method m_{entry} in this set, it finds the class c that declares this method and the top-level base class

Table 3.1 Calling convention of asynchronous calls

Top-level class	Run method	Start method
Runnable	run()	start()
AsyncTask	onPreExecute()	execute()
AsyncTask	doInBackground()	onPreExecute()
AsyncTask	onPostExecute()	doInBackground()
Message	handleMessage()	sendMessage()

```
public class AsyncTask{
    public AsyncTask execute(Params... params){
        executeStub(params);
    }
    public AsyncTask executeStub(Params...params){
        onPreExecute();
        Result result = doInBackground(params);
        onPostExecuteStub(result);
    }
    public void onPostExecuteStub(Result result){
        onPostExecute(result);
    }
}
```

Fig. 3.5 Stub code for dataflow of AsyncTask.execute

base that c inherits from. Then, it searches the pair of *base* and m_{entry} in the CM_{async} set. If a match is found, the method m_{entry} is a "callee" by convention. The algorithm thus looks up m_{entry} in the map SR_{async} to find the corresponding "caller" m_{start}. Each call to m_{start} is further examined and a points-to analysis is performed on the "this" reference making the call. If class c of method m_{entry} belongs to the points-to set, we can ensure the calling relationship between the caller m_{start} and the callee m_{entry} and remove the callee from the entry point set.

To indicate the data dependency between these two methods, we introduce a stub which connects the parameters of the asynchronous call to the corresponding parameters of its callee. Figure 3.5 depicts the example stub code for AsyncTask, where the parameter of execute() is first passed to doInBackground() through the stub executeStub(), and then the return from this asynchronous execution is further transferred to onPostExecute() via onPostExecuteStub().

Once the algorithm has reduced the number of entry point methods in M_{entry}, we explore all code reachable from those entry points, including both synchronous and asynchronous calls. We further determine the user interactivity of an entry point by examining its top-level base class. If the entry point callback overrides a counterpart declared in one of the three top-level UI-related interfaces (i.e., android.graphics.drawable.Drawable.Callback, android.view.accessibi- lity.AccessibilityEventSource, and android.- view.KeyEvent.Callback), we then consider the derived entry point method as a user interface.

3.3.4.2 Constant Analysis

We conduct constant analysis for any critical parameters of security sensitive API calls. These calls may expose security-related behaviors depending upon the values of their constant parameters. For example, `Runtime.exec()` can directly execute shell commands, and file or database operations can interact with distinctive targets by providing the proper URIs as input parameters.

To understand these semantic-level differences, we perform backward dataflow analysis on selected parameters and collect all possible constant values on the backward trace. We generate a constant set for each critical API argument and mark the parameter as "Constant" in the corresponding node on the WC-ADG. While a more complete string constant analysis is also possible, the computation of regular expressions is fairly expensive for static analysis. The substring set currently generated effectively reflects the semantics of a critical API call and is sufficient for further feature extraction.

3.3.4.3 API Dependency Construction

We perform global dataflow analysis to discover data dependencies between API nodes and build the edges on WC-ADG. However, it is very expensive to analyze every single API call made by an app. To address computational efficiency and our interests on security analysis, we choose to analyze only the security-related API calls. Permissions are strong indicators of security sensitivity in Android systems, so we leverage the API-permission mapping from PScout [18] to focus on permission-related API calls.

Our static dataflow analysis is similar to the "split"-based approach used by CHEX [16]. Each program split includes all code reachable from a single entry point. Dataflow analysis is performed on each split, and then cross-split dataflows are examined. The difference between our analysis and that of CHEX lies in the fact that we compute larger splits due to the consideration of asynchronous calling conventions.

We make a special consideration for reflective calls within our analysis. In Android programs, reflection is realized by calling the method `java.lang.reflect.Method.invoke()`. The "this" reference of this API call is a `Method` object, which is usually obtained by invoking either `getMethod()` or `getDeclaredMethod()` from `java.lang.Class`. The class is often acquired in a reflective manner too, through `Class.forName()`. This API call resolves a string input and retrieves the associated `Class` object.

We consider any reflective `invoke()` call as a sink and conduct backward dataflow analysis to find any prior data dependencies. If such an analysis reaches string constants, we are able to statically resolve the class and method information. Otherwise, the reflective call is not statically resolvable. However, statically unresolvable behavior is still represented within the WC-ADG as nodes which contain no constant parameters. Instead, this reflective call may have several preceding APIs, from a dataflow perspective, which are the sources of its metadata.

3.4 Android Malware Classification

We generate WC-ADGs for both benign and malicious apps. Each unique graph is associated with a feature that we use to classify Android malware and benign applications.

3.4.1 Graph Matching Score

To quantify the similarity of two graphs, we first compute a graph edit distance. To our knowledge, all existing graph edit distance algorithms treat node and edge uniformly. However, in our case, our graph edit distance calculation must take into account the different weights of different API nodes. At present, we do not consider assigning different weights on edges because this would lead to prohibitively high complexity in graph matching. Moreover, to emphasize the differences between two nodes in different labels, we do not seek to relabel them. Instead, we delete the old node and insert the new one subsequently. This is because node "relabeling" cost, in our context, is not the string edit distance between the API labels of two nodes. It is the cost of deleting the old node plus that of adding the new node.

Definition 2. The *Weighted Graph Edit Distance* (WGED) of two Weighted Contextual API Dependency Graphs G and G', with a uniform weight function β, is the minimum cost to transform G to G':

$$wged(G, G', \beta) = min(\sum_{v_I \in \{V'-V\}} \beta(v_I) + \sum_{v_D \in \{V-V'\}} \beta(v_D) + |E_I| + |E_D|), \quad (3.1)$$

where V and V' are respectively the vertices of two graphs, v_I and v_D are individual vertices inserted to and deleted from G, while E_I and E_D are the edges added to and removed from G.

WGED presents the absolute difference between two graphs. This implies that $wged(G, G')$ is roughly proportional to the sum of graph sizes and therefore two larger graphs are likely to be more distant to one another. To eliminate this bias, we normalize the resulting distance and further define *Weighted Graph Similarity* based upon it.

Definition 3. The *Weighted Graph Similarity* of two Weighted Contextual API Dependency Graphs G and G', with a weight function β, is,

$$wgs(G, G', \beta) = 1 - \frac{wged(G, G', \beta)}{wged(G, \emptyset, \beta) + wged(\emptyset, G', \beta)}, \quad (3.2)$$

where $\emptyset$ is an empty graph. $wged(G, \emptyset, \beta) + wged(\emptyset, G', \beta)$ then equates the maximum possible edit cost to transform G to G'.

3.4.2   Weight Assignment

Instead of manually specifying the weights on different APIs (in combination of their attributes), we wish to see a near-optimal weight assignment.

3.4.2.1   Selection of Critical API Labels

Given a large number of API labels (unique combinations of API names and attributes), it is unrealistic to automatically assign weights for every one of them. Our goal is malware classification, so we concentrate on assigning weights to labels for the security-sensitive APIs and critical combinations of their attributes. To this end, we perform *concept learning* to discover critical API labels. Given a positive example set (PES) containing malware graphs and a negative example set (NES) containing benign graphs, we seek a critical API label (CA) based on two requirements: (1) frequency(CA,PES) > frequency(CA,NES) and (2) frequency(CA,NES) is less than the median frequency of all critical API labels in NES. The first requirement guarantees that a critical API label is more sensitive to a malware sample than a benign one, while the second requirement ensures the infrequent presence of such an API label in the benign set. Consequently, we have selected 108 critical API labels. Our goal becomes the assignment of appropriate weights to these 108 labels while assigning a default weight of 1 to all remaining API labels.

3.4.2.2   Weight Assignment

Intuitively, if two graphs come from the same malware family and share one or more critical API labels, we must maximize the similarity between the two. We call such a pair of graphs a "homogeneous pair". Conversely, if one graph is malicious and the other is benign, even if they share one or more critical API labels, we must minimize the similarity between the two. We call such a pair of graphs a "heterogeneous pair". Therefore, we cast the problem of weight assignment to be an optimization problem.

Definition 4. The *Weight Assignment* is an optimization problem to maximize the result of an objective function for a given set of graph pairs $\{< G, G' >\}$:

$$\max f(\{< G, G' >\}, \beta) = \sum_{\substack{<G,G'> \text{ is a} \\ homogeneous\ pair}} wgs(G, G', \beta) - \sum_{\substack{<G,G'> \text{ is a} \\ heterogeneous\ pair}} wgs(G, G', \beta)$$

s.t.   (3.3)

$$1 \leq \beta(v) \leq \theta, \textit{if } v \textit{ is a critical node};$$

$$\beta(v) = 1, \textit{otherwise},$$

Fig. 3.6 A feedback loop to solve the optimization problem

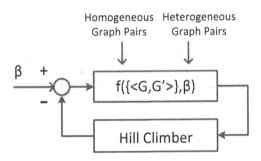

where β is the weight function that requires optimization; θ is the upper bound of a weight. Empirically, we set θ to be 20.

To achieve the optimization of Eq. (3.3), we use the *Hill Climbing* algorithm [19] to implement a feedback loop that gradually improves the quality of weight assignment. Figure 3.6 presents such a system, which takes two sets of graph pairs and an initial weight function β as inputs. β is a discrete function which is represented as a weight vector. At each iteration, Hill Climbing adjusts a single element in the weight vector and determines whether the change improves the value of objective function $f(\{< G, G' >\}, \beta)$. Any change that improves $f(\{< G, G' >\}, \beta)$ is accepted, and the process continues until no change can be found that further improves the value.

3.4.3 Implementation and Graph Database Query

To compute the weighted graph similarity, we use a bipartite graph matching tool [13]. We cannot directly use this graph matching tool because it does not support assigning different weights on different nodes in a graph. To work around this limitation, we enhanced the bipartite algorithm to support weights on individual nodes.

We then develop a graph database, where dependency graphs are stored into multiple buckets. Each bucket is labeled according to the presence of critical APIs. To ensure the scalability, we implement the bucket-based indexing with a hash map where the key is the API package bitvector and the value is a corresponding graph set. Empirically, we found this one-level indexing efficient enough for our problem. If the database grows much larger, we can transition to a hierarchical database structure, such as vantage point tree [20], under each bucket.

3.4.4 Malware Classification

3.4.4.1 Anomaly Detection

We have implemented a detector to conduct anomaly detection. Given an app, the detector provides a binary result that indicates whether the app is abnormal or not. To achieve this goal, we build a graph database for benign apps. The detector then attempts to match the WC-ADGs of the given app against the ones in database. If a sufficiently similar one for any of the behavior graphs is not found, an anomaly is reported by the detector. We have set the similarity threshold to be 70 % per our empirical studies.

3.4.4.2 Signature Detection

We next use a classifier to perform signature detection. Our signature detector is a multi-label classifier designed to identify the malware families of unknown malware instances.

To enable classification, we first build a malware graph database. To this end, we conduct static analysis on the malware samples from the Android Malware Genome Project [21, 22] to extract WC-ADGs. In order to consider only the unique graphs, we remove any graphs that have a high level of similarity to existing ones. With experimental study, we consider a high similarity to be greater than 80 %. Further, to guarantee the distinctiveness of malware behaviors, we compare these malware graphs against our benign graph set and remove the common ones.

Next, given an app, we generate its feature vector for classification purpose. In such a vector, each element is associated with a graph in our database. And, in turn, all the existing graphs are projected to a feature vector. In other words, there exists a one-to-one correspondence between the elements in a feature vector and the existing graphs in the database. To construct the feature vector of the given app, we produce its WC-ADGs and then query the graph database for all the generated graphs. For each query, a best matching graph is found. The similarity score is then put into the feature vector at the position corresponding to this best matching graph. Specifically, the feature vector of a known malware sample is attached with its family label so that the classifier can understand the discrepancy between different malware families.

Figure 3.7 gives an example of feature vectors. In our malware graph database of 862 graphs, a feature vector of 862 elements is constructed for each app. The two behavior graphs of ADRD are most similar to graph G6 and G7, respectively, from the database. The corresponding elements of the feature vector are set to the similarity scores of those features. The rest of the elements remain set to zero.

Once we have produced the feature vectors for the training samples, we can next use them to train a classifier. We select Naïve Bayes algorithm for the classification. In fact, we can choose different algorithms for the same purpose. However, since our graph-based features are fairly strong, even Naïve Bayes can produce satisfying

	G1	G2	G3	G4	G5	G6	G7	G8	...	G861	G862
ADRD	0	0	0	0	0	0.8	0.9	0	...	0	0
DroidDream	0.9	0	0	0	0.8	0.7	0.7	0	...	0	0
DroidKungFu	0	0.7	0	0	0.6	0	0.6	0	...	0	0.9

Fig. 3.7 An example of feature vectors

results. Naïve Bayes also has several advantages: it requires only a small amount of training data; parameter adjustment is straightforward and simple; and runtime performance is favorable.

3.5 Evaluation

3.5.1 Dataset and Experiment Setup

We collected 2200 malware samples from the Android Malware Genome Project [21] and McAfee, Inc, a leading antivirus company. To build a benign dataset, we received a number of benign samples from McAfee, and we downloaded a variety of popular apps having a high ranking from Google Play. Specifically, without loss of generality, we followed the prior approach in major research work [6, 23] and collected the top 1000 apps from each of the 16 categories. To further sanitize this benign dataset, we sent these apps to the VirusTotal service for inspection. The final benign dataset consisted of 13,500 samples. We performed the behavior graph generation, graph database creation, graph similarity query and feature vector extraction using this dataset. We conducted the experiment on a test machine equipped with Intel(R) Xeon(R) E5-2650 CPU (20 M Cache, 2 GHz) and 128 GB of physical memory. The operating system is Ubuntu 12.04.3 (64bit).

3.5.2 Summary of Graph Generation

We summarize the characteristics of the behavior graphs generated from both benign and malicious apps. Figures 3.8 and 3.9 illustrate the distribution of the number of graphs generated from benign and malicious apps. On average, 7.8 graphs are computed from each benign app, while 9.8 graphs are generated from each malware instance. Most apps focus on limited functionalities and do not produce a large number of behavior graphs. In 92 % of benign samples and 98 % of malicious ones, no more than 20 graphs are produced from an individual app.

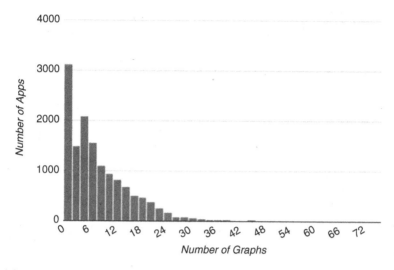

Fig. 3.8 Distribution of the number of graphs in benign apps

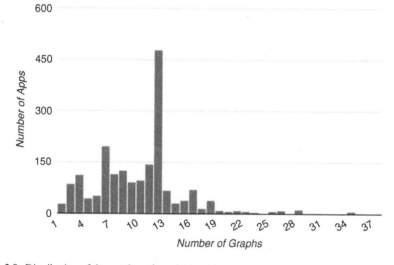

Fig. 3.9 Distribution of the number of graphs in malware

Figures 3.10 and 3.11 present the distribution of the number of nodes in benign and malicious behavior graphs. A benign graph, on average, has 15 nodes, while a malicious graph carries 16.4. Again, most of the activities are not intensive, so the majority of these graphs have a small number of nodes. Statistics show that 94 % of the benign graphs and 91 % of the malicious ones carry less than 50 nodes. These facts serve as the basic requirements for the scalability of our approach, since the runtime performance of graph matching and query is largely dependent upon the number of nodes and graphs, respectively.

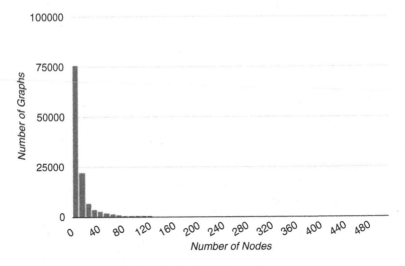

Fig. 3.10 Distribution of the number of nodes in benign graphs

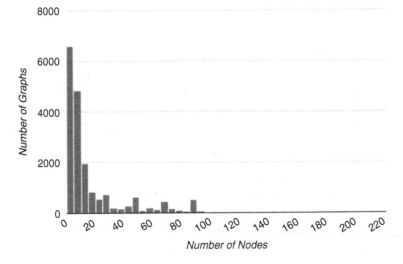

Fig. 3.11 Distribution of the number of nodes in malware graphs

3.5.3 Classification Results

3.5.3.1 Signature Detection

We use a multi-label classification to identify the malware family of the unrec-
ognized malicious samples. Therefore, we expect to only include those malware
behavior graphs, that are well labeled with family information, into the database.
To this end, we rely on the malware samples from the Android Malware Genome

Project and use them to construct the malware graph database. Consequently, we built such a database of 862 unique behavior graphs, with each graph labeled with a specific malware family.

We then selected 1050 malware samples from the Android Malware Genome Project and used them as a training set. Next, we would like to collect testing samples from the rest of our collection. However, a majority of malware samples from McAfee are not strongly labeled. Over 90 % of the samples are coarsely labeled as "Trojan" or "Downloader", but in fact belong to a specific malware family (e.g., DroidDream). Moreover, even VirusTotal cannot provide reliable malware family information for a given sample because the antivirus products used by VirusTotal seldom reach a consensus. This fact tells us two things: (1) It is a non-trivial task to collect evident samples as the ground truth in the context of multi-label classification; (2) multi-label malware detection or classification is, in general, a challenging real-world problem.

Despite the difficulty, we obtained 193 samples, each of which is detected as the same malware by major AVs. We then used those samples as testing data. The experiment result shows that our classifier can correctly label 93 % of these malware instances.

Among the successfully labeled malware samples are two types of Zitmo variants. One uses HTTP for communication, and the other uses SMS. While the former one is present in our malware database, the latter one was not. Nevertheless, our signature detector is still able to capture this variant. This indicates that our similarity metrics effectively tolerate variations in behavior graphs.

We further examined the 7 % of the samples that were mislabeled. It turns out that the mislabeled cases can be roughly put into two categories. First, DroidDream samples are labeled as DroidKungFu. DroidDream and DroidKungFu share multiple malicious behaviors such as gathering privacy-related information and hidden network I/O. Consequently, there exists a significant overlap between their WC-ADGs. Second, Zitmo, Zsone and YZHC instances are labeled as one another. These three families are SMS Trojans. Though their behaviors are slightly different from each other, they all exploit `sendTextMessage()` to deliver the user's information to an attacker-specified phone number. Despite the mislabeled cases, we still manage to successfully label 93 % of the malware samples with a Naïve Bayes classifier. Applying a more advanced classification algorithm will further improve the accuracy.

3.5.3.2 Anomaly Detection

Since we wish to perform anomaly detection using our benign graph database, the coverage of this database is essential. In theory, the more benign apps that the database collects, the more benign behaviors it covers. However, in practice, it is extremely difficult to retrieve benign apps exhaustively. Luckily, different benign apps may share the same behaviors. Therefore, we can focus on unique behaviors (rather than unique apps). Moreover, with more and more apps being fed into the

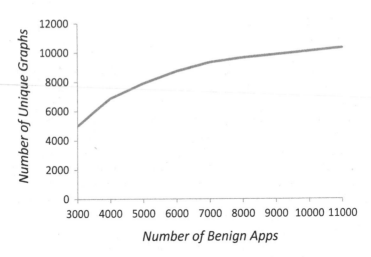

Fig. 3.12 Convergence of unique graphs in benign apps

benign database, the database size grows slower and slower. Figure 3.12 depicts our discovery. When the number of apps increases from 3000 to 4000, there is a sharp increase (2087) of unique graphs. However, when the number of apps grows from 10,000 to 11,000, only 220 new, unique graphs are generated, and the curve begins to flatten.

We built a database of 10,420 unique graphs from 11,400 benign apps. Then, we tested 2200 malware samples against the benign classifier. The false negative rate was 2 %, which indicates that 42 malware instances were not detected. However, we noted that most of the missed samples are exploits or Downloaders. In these cases, their bytecode programs do not bear significant API-level behaviors, and therefore generated WC-ADGs do not necessarily look abnormal when compared to benign ones. At this point, we have only considered the presence of constant parameters in an API call. We did not further differentiate API behaviors based upon constant values. Therefore, we cannot distinguish the behaviors of Runtime.exec() calls or network I/O APIs with varied string inputs. Nevertheless, if we create a custom filter for these string constants, we will then be able to identify these malware samples and the false negative rate will drop to 0.

Next, we used the remaining 2100 benign apps as test samples to evaluate the false positive rate of our anomaly detector. The result shows that 5.15 % of clean apps are mistakenly recognized as suspicious ones during anomaly detection. This means, if our anomaly detector is applied to Google Play, among the approximately 1200 new apps per day [24], around 60 apps will be mislabeled as containing anomalies and be bounced back to the developers. We believe that this is an acceptable ratio for vetting purpose. Moreover, since we do not reject the suspicious apps immediately, but rather ask the developers for justifications instead, we can further eliminate these false positives during this interactive process. In addition, as we add more benign samples into the dataset, the false positive rate will further decrease.

Further, we would like to evaluate our anomaly detector with malicious samples from new malware families. To this end, we retrieve a new piece of malware called Android.HeHe, which was first found and reported in January 2014 [25]. Android.HeHe exercises a variety of malicious functionalities such as SMS interception, information stealing and command-and-control. This new malware family does not appear in the samples from the Android Malware Genome Project, which were collected from August 2010 to October 2011, and therefore cannot be labeled via signature detection. DroidSIFT generates 49 WC-ADGs for this sample and, once these graphs are presented to the anomaly detector, a warning is raised indicating the abnormal behaviors expressed by these graphs.

3.5.3.3 Detection of Transformation Attacks

We collected 23 DroidDream samples, which are all intentionally obfuscated using a transformation technique [4], and 2 benign apps that are deliberately disguised as malware instances by applying the same technique. We ran these samples through our anomaly detection engine and then sent the detected abnormal ones through the signature detector. The result shows that while 23 true malware instances are flagged as abnormal ones in anomaly detection, the 2 clean ones also correctly pass the detection without raising any warnings. We then compared our signature detection results with antivirus products. To obtain detection results of antivirus software, we sent these samples to VirusTotal and selected 10 anti-virus (AV) products (i.e., AegisLab, F-Prot, ESET-NOD32, DrWeb, AntiVir, CAT-QuickHeal, Sophos, F-Secure, Avast, and Ad-Aware) that bear the *highest* detection rates. Notice that we consider the detection to be successful only if the AV can correctly flag a piece of malware as DroidDream or its variant. In fact, to our observation, many AV products can provide partial detection results based upon the native exploit code included in the app package or common network I/O behaviors. As a result, they usually recognize these DroidDream samples as "exploits" or "Downloaders" while missing many other important malicious behaviors. Figure 3.13 presents the detection ratios of "DroidDream" across different detectors. While none of the antivirus products can achieve a detection rate higher than 61 %, DroidSIFT can successfully flag all the obfuscated samples as DroidDream instances. In addition, we also notice that AV2 produces a relatively high detection ratio (52.17 %), but it also mistakenly flags those two clean samples as malicious apps. Since the disguising technique simply renames the benign app package to the one commonly used by DroidDream (and thus confuses this AV detector), such false positives again explain that external symptoms are not robust and reliable features for malware detection.

Fig. 3.13 Detection ratio for obfuscated malware

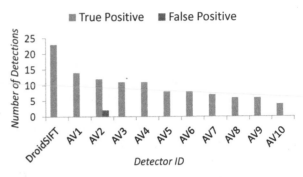

3.5.4   Runtime Performance

The average detection runtime of 3000 apps is 175.8 s, while the detection for a majority (86%) of apps is completed within 5 min. Further, most of the apps (96%) are processed within 10 min. The time cost of graph generation dominates the overall runtime, taking up at least 50% of total runtime for 83.5% of the apps. On the other hand, the signature and anomaly detectors are usually (i.e., in 98% of the cases) able to finish running in 3 and 1 min, respectively.

3.5.5   Effectiveness of Weight Generation and Weighted Graph Matching

Finally, we evaluated the effectiveness of the generated weights and weighted graph matching. Our weight generation automatically assign weights to the critical API labels, based on a training set of *homogeneous graph pairs* and *heterogeneous graph pairs*. Consequently, killProcess(), getMemoryInfo() and sendTextMessage() with a constant phone number, for example, are assigned with fairly high weights.

Then, given a graph pair sharing the same critical API labels, other than the pairs used for training, we want to compare their weighted graph similarity with the similarity score calculated by the standard bipartite algorithm. To this end, we randomly picked 250 homogeneous pairs and 250 heterogeneous pairs.

The results of these comparisons, presented in Figs. 3.14 and 3.15, conform to our expectation. Figure 3.14 shows that for every homogeneous pair, the similarity score generated by weighted graph matching is almost always higher than the corresponding one computed using standard bipartite algorithm. In addition, the bipartite algorithm sometimes produces an extremely low similarity (i.e., near zero) between two malicious graphs of the same family, while weighted graph matching manages to improve the similarity score significantly for these cases.

Fig. 3.14 Similarity between malicious graph pairs

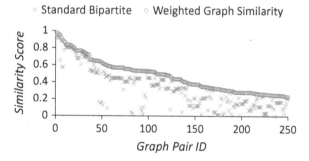

Fig. 3.15 Similarity between benign and malicious graphs

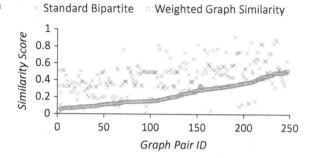

Similarly, Fig. 3.15 reveals that between a heterogeneous pair, the weighted similarity score is usually lower than the one from bipartite computation. Again, the bipartite algorithm occasionally considers a benign graph considerably similar to a malicious one, provided that they share the same API nodes. Such results can confuse a training system and the latter one thus fails to tell the differences between malicious and benign behaviors. On the other hand, weighted graph matching can effectively distinguish a malicious graph from a benign one, even if they both have the same critical API nodes.

We further attempted to implement the standard bipartite algorithm and apply it to our detectors. We then compared the consequent detection results with those of the detectors with weighted graph matching enabled. The results show that weighted graph matching significantly outperforms the bipartite one. While the signature detector using the former one correctly labels 93 % of malware samples, the detector with the latter one is able to only label 73 % of them. On the other hand, anomaly detection with the bipartite algorithm incurs a false negative rate of 10 %, which is 5 times greater than that introduced by the same detector using weighted matching.

The result indicates that our algorithm is more sensitive to critical API-level semantics than the standard bipartite graph matching, and thus can produce more reasonable similarity scores for the feature extraction.

References

1. McAfee Labs Threats report Fourth Quarter (2013) http://www.mcafee.com/us/resources/reports/rp-quarterly-threat-q4-2013.pdf
2. Zhou Y, Wang Z, Zhou W, Jiang X (2012) Hey, you, get off of my market: detecting malicious apps in official and alternative android markets. In: Proceedings of 19th annual network and distributed system security symposium (NDSS)
3. Grace M, Zhou Y, Zhang Q, Zou S, Jiang X (2012) RiskRanker: scalable and accurate zero-day android malware detection. In: Proceedings of the 10th international conference on mobile systems, applications and services (MobiSys)
4. Rastogi V, Chen Y, Jiang X (2013) DroidChameleon: evaluating android anti-malware against transformation attacks. In: Proceedings of the 8th ACM symposium on information, computer and communications security (ASIACCS)
5. Peng H, Gates C, Sarma B, Li N, Qi Y, Potharaju R, Nita-Rotaru C, Molloy I (2012) Using probabilistic generative models for ranking risks of android apps. In: Proceedings of the 2012 ACM conference on computer and communications security (CCS)
6. Aafer Y, Du W, Yin H (2013) DroidAPIMiner: mining API-level features for robust malware detection in android. In: Proceedings of the 9th international conference on security and privacy in communication networks (SecureComm)
7. Arp D, Spreitzenbarth M, Hübner M, Gascon H, Rieck K (2014) Drebin: efficient and explainable detection of android malware in your pocket. In: Proceedings of the 21th annual network and distributed system security symposium (NDSS)
8. Christodorescu M, Jha S, Seshia SA, Song D, Bryant RE (2005) Semantics-aware malware detection. In: Proceedings of the 2005 IEEE symposium on security and privacy (Oakland)
9. Fredrikson M, Jha S, Christodorescu M, Sailer R, Yan X (2010) Synthesizing near-optimal malware specifications from suspicious behaviors. In: Proceedings of the 2010 IEEE symposium on security and privacy (Oakland)
10. Kolbitsch C, Comparetti PM, Kruegel C, Kirda E, Zhou X, Wang X (2009) Effective and efficient malware detection at the end host. In: Proceedings of the 18th conference on USENIX security symposium
11. Chen KZ, Johnson N, D'Silva V, Dai S, MacNamara K, Magrino T, Wu EX, Rinard M, Song D (2013) Contextual policy enforcement in android applications with permission event graphs. In: Proceedings of the 20th annual network and distributed system security symposium (NDSS)
12. Soot: A Java Optimization Framework (2016) http://www.sable.mcgill.ca/soot/
13. Riesen K, Emmenegger S, Bunke H (2013) A novel software toolkit for graph edit distance computation. In: Proceedings of the 9th international workshop on graph based representations in pattern recognition
14. Lockheimer H (2012) Android and security. http://googlemobile.blogspot.com/2012/02/android-and-security.html
15. Oberheide J, Miller C (2012) Dissecting the android bouncer. In: SummerCon
16. Lu L, Li Z, Wu Z, Lee W, Jiang G (2012) CHEX: statically vetting android apps for component hijacking vulnerabilities. In: Proceedings of the 2012 ACM conference on computer and communications security (CCS)
17. Octeau D, McDaniel P, Jha S, Bartel A, Bodden E, Klein J, Traon YL (2013) Effective inter-component communication mapping in android with epicc: an essential step towards holistic security analysis. In: Proceedings of the 22nd USENIX security symposium
18. Au KWY, Zhou YF, Huang Z, Lie D (2012) PScout: analyzing the android permission specification. In: Proceedings of the 2012 ACM conference on computer and communications security (CCS)
19. Russell SJ, Norvig P (2013) Artificial intelligence: a modern approach, 3rd edn. Prentice-Hall, Inc., Upper Saddle River
20. Hu X, Chiueh TC, Shin KG (2009) Large-scale malware indexing using function-call graphs. In: Proceedings of the 16th ACM conference on computer and communications security (CCS)

21. Android Malware Genome Project (2012) http://www.malgenomeproject.org/
22. Zhou Y, Jiang X (2012) Dissecting android malware: characterization and evolution. In: Proceedings of the 33rd IEEE symposium on security and privacy. Oakland
23. Enck W, Octeau D, McDaniel P, Chaudhuri S (2011) A study of android application security. In: Proceedings of the 20th USENIX security symposium
24. Number of Android Applications (2014) http://www.appbrain.com/stats/number-of-android-apps
25. Dharmdasani H (2014) Android.HeHe: malware now disconnects phone calls. http://www.fireeye.com/blog/technical/2014/01/android-hehe-malware-now-disconnects-phone-calls.html

Chapter 4
Automatic Generation of Vulnerability-Specific Patches for Preventing Component Hijacking Attacks

Abstract *Component hijacking* is a class of vulnerabilities commonly appearing in Android applications. When these vulnerabilities are triggered by attackers, the vulnerable apps can exfiltrate sensitive information and compromise the data integrity on Android devices, on behalf of the attackers. It is often unrealistic to purely rely on developers to fix these vulnerabilities for two reasons: (1) it is a time-consuming process for the developers to confirm each vulnerability and release a patch for it; and (2) the developers may not be experienced enough to properly fix the problem. In this paper, we propose a technique for automatic patch generation. Given a vulnerable Android app (without source code) and a discovered component hijacking vulnerability, we automatically generate a patch to disable this vulnerability. We have implemented a prototype called *AppSealer* and evaluated its efficacy on apps with component hijacking vulnerabilities. Our evaluation on 16 real-world vulnerable Android apps demonstrates that the generated patches can effectively track and mitigate component hijacking vulnerabilities. Moreover, after going through a series of optimizations, the patch code only represents a small portion (15.9 % on average) of the entire program. The runtime overhead introduced by AppSealer is also minimal, merely 2 % on average.

4.1 Introduction

With the boom of Android devices, the security threats in Android are also increasing. Although the permission-based sandboxing mechanism enforced in Android can effectively confine each app's behaviors by only allowing the ones granted with corresponding permissions, a vulnerable app with certain critical permissions can perform security-sensitive behaviors on behalf of a malicious app. It is so called confused deputy attack. This kind of security vulnerabilities can present in numerous forms, such as privilege escalation [1], capability leaks [2], permission re-delegation [3], content leaks and pollution [4], component hijacking [5], etc.

Prior work primarily focused on automatic discovery of these vulnerabilities. Once a vulnerability is discovered, it can be reported to the developer and a patch is expected. Some patches can be as simple as placing a permission validation at the entry point of an exposed interface (to defeat privilege escalation [1] and permission

© The Author(s) 2016
M. Zhang, H. Yin, *Android Application Security*, SpringerBriefs in Computer Science, DOI 10.1007/978-3-319-47812-8_4

re-delegation [3] attacks), or withholding the public access to the internal data repositories (to defend against content leaks and pollution [4]), the fixes to the other problems may not be so straightforward.

In particular, component hijacking may fall into the latter category. When receiving a manipulated input from a malicious Android app, an app with a component hijacking vulnerability may exfiltrate sensitive information or tamper with the sensitive data in a critical data repository on behalf of the malicious app. In other words, a dangerous information flow may happen in either an outbound or inbound direction depending on certain external conditions and/or the internal program state.

A prior effort has been made to perform static analysis to discover *potential* component hijacking vulnerabilities [5]. Static analysis is known to be conservative in nature and may raise false positives. To name a few, static analysis may find a viable execution path for information flow, which may never be reached in actual program execution; static analysis may find that interesting information is stored in some elements in a database, and thus has to conservatively treat the entire database as sensitive. Upon receiving a discovered vulnerability, the developer has to manually confirm if the reported vulnerability is real. It may also be nontrivial for the (often inexperienced) developer to properly fix the vulnerability and release a patch for it. As a result, these discovered vulnerabilities may not be addressed for long time or not addressed at all, leaving a big time window for attackers to exploit these vulnerabilities. Out of the 16 apps with component hijacking vulnerabilities we tested, only 3 of them were fixed in 1 year.

To close this window, we aim to automatically generate a patch that is specific to the discovered component hijacking vulnerability. In other words, we would like to automatically generate a *vulnerability-specific* patch on the original program to block the vulnerability as a whole, not just a set of malicious requests that exploit the vulnerability.

While automatic patch generation is fairly new in the context of Android applications, a great deal of research has been done for traditional client and server programs (with and without source code). Many efforts have been made to automatically generate signatures to block bad inputs, by performing symbolic execution and path exploration [6–10]. This approach is effective if the vulnerability is triggered purely by the external input and the library functions can be easily modeled in symbolic execution. However, for android applications, due to the asynchronous nature of the program execution, a successful exploitation may depend on not only the external input, but also the system events and API-call return values. Other efforts have been made to automatically generate code patches within the vulnerable programs to mitigate certain kinds of vulnerabilities. To name a few, AutoPaG [11] focused on buffer overflow and general boundary errors; IntPatch [12] addressed integer-overflow-to-buffer-overflow problem; To defeat zero-day worms, Sidiroglou and Keromytis [13] proposed an end-point first-reaction mechanism that tries to automatically patch vulnerable software by identifying and transforming the code surrounding the exploited software flaw; and VSEF [14] monitors and instruments the part of program execution relevant to specific vulnerability, and creates execution-based filters which filter out exploits based on vulnerable program's execution trace, rather than solely upon input string.

In principle, our automatic patch generation technique falls into the second category. However, our technique is very different from these existing techniques, because it requires a new machinery to address this new class of vulnerabilities. The key of our patch generation technique is to place *minimally* required code into the vulnerable program to *accurately* keep track of dangerous information originated from the exposed interfaces and effectively block the attack at the security-sensitive APIs.

To achieve this goal, we first perform static bytecode analysis to identify small but complete *program slices* that lead to the discovered vulnerability. Then we devise several shadowing mechanisms to insert new variables and instructions along the program slices, for the purpose of keeping track of dangerous information at run-time. In the end, we apply a series of optimizations to remove redundant instructions to minimize the footprint of the generated patch. Consequently, the automatically generated patch can be guaranteed to completely disable the vulnerability with minimal impact on runtime performance and usability.

We implement a prototype, *AppSealer*, in 16 thousand lines of Java code, based on the Java bytecode optimization framework Soot [15]. We leverage Soot's capability to perform static dataflow analysis and bytecode instrumentation. We evaluate our tool on 16 real-world Android apps with component hijacking vulnerabilities. Our experiments show that the patched programs run correctly, while the vulnerabilities are effectively mitigated.

4.2 Problem Statement and Approach Overview

4.2.1 Running Example

Figure 4.1 presents a synthetic running example in Java source code, which has a component hijacking vulnerability. More concretely, the example class is one of the Activity components in an Android application. It extends an Android Activity and overrides several callbacks including onCreate(), onStart() and onDestroy().

Upon receiving a triggering Intent, the Android framework creates the Activity and further invokes these callbacks. Once the Activity is created, onCreate() retrieves the piggybacked URL string from the triggering Intent. Next, it saves this URL to an instance field addr if the resulting string is not null, or uses the DEFAULT_ADDR otherwise. When the Activity starts, onStart() method acquires the latest location information by calling getLastKnownLocation(), and stores it to a static field location. Further, onDestroy() reads the location object from this static field, encodes the data into a string, encrypts the byte array of the string and sends it to the URL specified by addr through a raw socket.

```
1  public class VulActivity extends Activity{
2    private String DEFAULT_ADDR = "http://default.url";
3    private byte DEFAULT_KEY = 127;
4
5    private String addr;
6    private static Location location;
7    private byte key;
8
9    /* Entry point of this Activity */
10   public void onCreate(Bundle savedInstanceState){
11     this.key = DEFAULT_KEY;
12
13     this.addr = getIntent().getExtras().getString("url");
14     if(this.addr == null){
15       this.addr = DEFAULT_ADDR;
16     }
17   }
18
19   public void onStart(){
20     VulActivity.location = getLocation();
21   }
22
23   public void onDestroy(){
24     String location =
25       Double.toString(VulActivity.location.getLongitude()) + "," + Double.
         toString(VulActivity.location.getLatitude());
26     byte[] bytes = location.getBytes();
27     for(int i=0; i<bytes.length; i++)
28       bytes[i] = crypt(bytes[i]);
29     String url = this.addr;
30     post(url, bytes);
31   }
32
33   public byte crypt(byte plain){
34     return (byte)(plain ^ key);
35   }
36
37   public Location getLocation(){
38     Location location = null;
39     LocationManager locationManager = (LocationManager)getSystemService(Context
         .LOCATION_SERVICE);
40     location = locationManager.getLastKnownLocation(LocationManager.
         GPS_PROVIDER);
41     return location;
42   }
43
44   public void post(String addr, byte[] bytes){
45     URL url = new URL(addr);
46     HttpURLConnection conn = (HttpURLConnection)url.openConnection();
47     ...
48     OutputStream output = conn.getOutputStream();
49     output.write(bytes, 0, bytes.length);
50     ...
51   }
52 }
```

Fig. 4.1 Java code for the running example

This program is subject to component hijacking attack, because a malicious app may send an Intent to this Activity with a URL specified by the attacker. As a result, this vulnerable app will send the location information to the attacker's URL. In other words, this vulnerability allows the malicious app to retrieve sensitive information without having to declare the related permissions (i.e., `android.permission.ACCESS_FINE_LOCATION` and `android.permission.INTERNET`).

Besides information leakage, a component hijacking attack may happen in the reverse direction. That is, a vulnerable program may allow a malicious app to modify the content of certain sensitive data storage, such as the Contacts database, even though the malicious app is not granted with the related permissions.

4.2.2 Problem Statement

We anticipate our proposed technique to be deployed as a security service in the Android marketplace, as illustrated in Fig. 4.2. Both the existing apps and the newly submitted apps must go through the vetting process by using static analysis tools like CHEX [5]. If a component hijacking vulnerability is discovered in an app, its developer will be notified, and a patch will be automatically generated to disable the discovered vulnerability. So the vulnerable apps will never reach the end users. This approach wins time for the developer to come up with a more fundamental solution to the discovered security problem. Even if the developer does not have enough skills to fix the problem or is not willing to, the automatically generated patch can serve as a permanent solution for most cases (if not all).

In addition, it is also possible to deploy our technique with more ad-hoc schemes. For instance, an enterprise can maintain its private app repository and security service too. The enterprise service conducts vetting and necessary patching before an app gets into the internal app pool, and thus employees are protected from vulnerable apps. Alternatively, establishing third-party public services and repositories is also viable and can benefit end users.

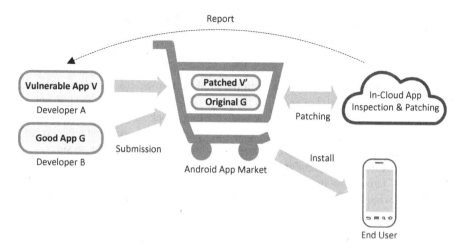

Fig. 4.2 Deployment of AppSealer

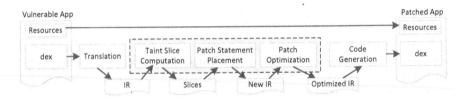

Fig. 4.3 Architecture of AppSealer

4.2.3 Approach Overview

Figure 4.3 depicts the workflow of our automatic patch generation technique. It takes the following steps:

(1) **IR Translation.** An Android app generally consists of a Dalvik bytecode executable file, manifest files, native libraries, and other resources. Our patch generation is performed on the Dalvik bytecode program. So the other files remain the same, and will be repackaged into the new app in the last step. To facilitate the subsequent static analysis, code insertion, and code optimization, we first translate Dalvik bytecode into an Intermediate Representation (IR). In particular, we first convert the DEX into Java bytecode program using dex2jar [16], and then using Soot [15], translate the Java bytecode into Jimple IR.

(2) **Slice Computation.** On Jimple IR, we perform *flow-sensitive context-sensitive inter-procedural* dataflow analysis to discover component hijacking flows. We track the propagation of certain "tainted" sensitive information from sources like internal data storage and exposed interfaces, and detect if the tainted information propagates into the dangerous data sinks. By performing both forward dataflow analysis and backward slicing, we compute one or more program slices that directly contribute to the dangerous information flow. To distinguish with other kinds of program slices, we call this slice *taint slice*, as it includes only the program statements that are involved in the taint propagation from a source to a sink.

(3) **Patch Statement Placement.** With the guidance of the computed taint slices, we place shadow statements into the IR program. The inserted shadow state-ments serve as runtime checks to actually keep track of the taint propagation while the Android application is running. In addition, guarding statements are also placed at the sinks to block dangerous information flow right on site.

(4) **Patch Statement Optimization.** We further optimize the inserted patch state-ments. This is to remove the redundant statements that are inserted from the previous step. As Soot's built-in optimizations do not apply well on these patch statements, we devise three custom optimization techniques to safely manipulate the statements and remove redundant ones. Thereafter, the optimized code is now amenable to the built-in optimizations. Consequently,

after going through both custom and built-in optimizations, the added patch statements can be reduced to the minimum, ensuring the best performance of the patched bytecode program.

(5) **Bytecode Generation.** At last, we convert the modified Jimple IR into a new package (.APK file). To do so, we translate the Jimple IR to Java package using Soot, and then re-target Java bytecode to Dalvik bytecode using Android SDK. In the end, we repackage the patched DEX file with old resources and create the new .APK file.

4.3 Taint Slice Computation

Taking a Jimple IR program and the sources and sinks specified in the security policies as input, our application-wide dataflow analysis will output one or more taint slices for dangerous information flows. Our analysis takes the following steps: (1) we locate the information sources in the IR program, and starting from each source, we perform forward dataflow analysis to generate a taint propagation graph; (2) if a corresponding sink is found in this taint propagation graph, we perform backward dependency analysis from the sink node and generate a taint slice.

We follow the similar approach elaborated in Chap. 3 to conduct dataflow analyses, and particularly we take special consideration of programming features in Android apps, such as static and instance fields, Intent, threads, etc.

4.3.1 Running Example

Figure 4.4 illustrates the taint slices for the running example, after using both forward dataflow analysis and backward dependency analysis. There are two slices in this graph, each one of which represents the data propagation of one source. The left branch describes the taint propagation of "gps" information, which originates from the invocation of getLastKnownLocation(). The data is then saved to a static field location, before a series of long-to-string and string-to-bytearray conversions in onDestroy(). Converted byte array is further passed to crypt() for byte-level encryption. The right branch begins with Intent receiving in onCreate(). The piggybacked "url" data is thus extracted from the Intent and stored into an instance field addr. The two branches converge at post(String,byte[]), when both the encrypted byte array and "addr" string are fed into the two parameters, respectively. "addr" string is used to construct an URL, then a connection, and further an OutputStream object, while the byte array serves as the payload. In the end, both sources flow into the sink OutputStream. write(byte[],int,int), which sends the payload to the designated address.

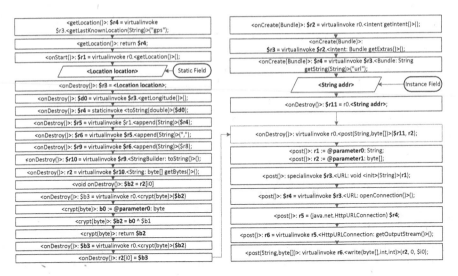

Fig. 4.4 Taint slices for the running example

4.4   Patch Statement Placement

Static dataflow analysis is usually conservative and may lead to false positives. Therefore, we insert instrumentation code in these taint propagation slices. The inserted code serves as runtime checks to actually keep track of the taint propagation while the Android application is running. The sinks are also instrumented to examine the taint and confine the information leakage.

We create shadows, for each data entity defined or used within the slices, to track its runtime taint status. The data entities outside the slices do not need to be shadowed, because they are irrelevant to the taint propagation. We create shadows for different types of data entities individually. Static or instance fields are shadowed by adding boolean fields into the same class definition. A local variable is shadowed with a boolean local variable within the same method scope, so that compiler optimizations can be applied smoothly on related instrument code.

Shadowing method parameters and return value requires special considerations. Firstly, the method prototype needs modification. Extra "shadow parameters" are added to parameter list to pass the shadows of actual parameters and return value from caller to callee. Secondly, parameters are passed-by-value in Java, and therefore primitive-typed local shadow variables (boolean) need to be wrapped as objects before passing to a callee. Otherwise, the change of shadows in the callee cannot be reflected in the caller. To this end, we define a new class `BoolWrapper`. This class only contains one boolean instance field, which holds the taint status, and a default constructor, as shown below. Notice that the Java `Boolean` class cannot serve the same purpose because it is treated the same as primitive boolean type once passed as parameter.

```
public class BoolWrapper extends java.lang.Object{
  public boolean b;
  public void <init>(){
    BoolWrapper r0;
    r0 := @this: BoolWrapper;
    specialinvoke r0.<java.lang.Object: void <init>()>();
    return; }
}
```

With the created shadows, we instrument sources, data propagation code and sinks. At the source of information flow, we introduce taint by setting the corresponding shadow variable to "true". For data propagation code, we instrument an individual instruction depending on its type. (1) If the instruction is an assignment statement or unary operation, we insert a definition statement on correlative shadow variables. (2) If it is a binary operation, a binary OR statement is inserted to operate on shadow variables. If one of the operators is a constant, we replace its shadow with a constant "false". (3) Or, if it is a function call, we need to add code to bind shadows of actual parameters and return value to shadow parameters. (4) Further, if the instruction is a API call, we model and instrument the API with its taint propagation logic.

We generally put APIs into the following categories and handle each category with a different model. "get" APIs have straightforward taint propagation logic, always propagating taint from parameters to their return values. Therefore, we generate a default rule, which propagates taint from any of the input parameters to the return value. Similarly, simple "set" APIs are modeled as they propagate taint from one parameter to another parameter or "this" reference. APIs like Vector.add(Object) inserts new elements into an aggregate construct and thus can be modeled as a binary operation, such that the object is tainted if it is already tainted or the newly added element is tainted. APIs like android.content.ContentValues.put(String key, Byte value) that operate on (key, value) pairs can have more precise handling. In this case, an element is stored and accessed according to a "key". To track taint more precisely, we keep a taint status for each key, so the taint for each (key, value) pair is updated individually.

Then, at the sink, we insert code before the sink API to check the taint status of the sensitive parameter. If it turns out the critical parameter is tainted, the inserted code will query a separate policy service app for decision and a warning dialog is then displayed to the user.

We also devise taint cleaning mechanism. That is, if a variable is redefined to be an untainted variable or a constant outside taint propagation slices, we thus insert a statement after that definition to set its shadow variable to 0 (false).

4.5 Patch Optimization

We further optimize the added instrumentation code. This is to remove the redundant bytecode instructions that are inserted from the previous step. As Soot's built-in optimizations do not apply well on this instrumentation code, we devise three custom optimization techniques to safely manipulate the instrumentation code and remove redundant ones. Thereafter, the optimized code is now amenable to the built-in optimizations. Consequently, after going through both custom and built-in optimizations, the added instrumentation code can be reduced to a minimum, ensuring the best performance of the rewritten bytecode program.

To be more specific, we have devised four steps of optimizations, described as follows.

- **O1**: In instrumentation, we add a shadow parameter for every single actual parameter and return value. However, some of them are redundant because they don't contribute to taint propagations. Therefore, we can remove the inserted code which uses solely these unnecessary shadow parameters.
- **O2**: Next, we remove redundant shadow parameters from parameter list and adjust method prototype. Consequently, instrumentation code, that is used to initialize or update the taint status of these shadow parameters, can also be eliminated.
- **O3**: Further, if inserted taint tracking code is independent from the control-flow logic of a method, we can lift the tainting code from the method to its callers. Thus, the taint propagation logic is inlined.
- **O4**: After custom optimizations, instrumentation code is amenable to Soot's built-in optimization, such as constant propagation, dead code elimination, etc.

4.5.1 Optimized Patch for Running Example

Figure 4.5 presents the optimized patch for the running example. For the sake of readability, we present the patch code in Java as a "diff" to the original program, even though this patch is actually generated on the Jimple IR level. Statements with numeric line numbers are from the original program, whereas those with special line number "**P**" are patch statements. The underlines mark either newly introduced code or modified parts from old statements.

We can see that a boolean variable "addr_s0_t" is created to shadow the instance field "addr", and another boolean variable "location_s1_t" is created to shadow the static field "location". Then in onCreate(), the shadow variable "addr_s0_t" is set to 1 (tainted) when the Activity is created upon an external Intent. Otherwise it will be set to 0 (untainted). The shadow variable "location_s1_t" is set to 1 inside onStart(), after getLocation() is called. Note that this initialization is originally placed inside getLocation() after Ln.40, and a BoolWrapper is created for the return value of getLocation(). After applying the inlining optimization(O3), this assignment is lifted into the caller function onStart() and the BoolWrapper variable is also removed.

Due to the optimizations, many patch statements placed for tracking tainted status have been removed. For example, the taint logic in crypt() has been lifted up to the body of onDestroy() and further optimized there. The tainted values

```
 1  public class VulActivity extends Activity{
        ...
 5      private String addr;
 P      public boolean addr_s0_t;
 6      private static Location location;
 P      public static boolean location_s1_t;
        ...
10      public void onCreate(Bundle savedInstanceState){
        ...
13        this.addr=getIntent().getExtras().getString("url");
 P        if(isExternalIntent()){
 P          this.addr_s0_t = true;
 P        }else{
 P          this.addr_s0_t = false;
 P        }
14        if(this.addr == null){
15          this.addr = DEFAULT_ADDR;
 P          this.addr_s0_t = false;
16        }
17      }
18
19      public void onStart(){
20        VulActivity.location = getLocation();
 P        VulActivity.location_s1_t = true;
21      }
22
23      public void onDestroy(){
        ...
29        String url = this.addr;
 P        BoolWrapper bytes_s1_w = new BoolWrapper();
 P        bytes_s1_w.b = VulActivity.location_s1_t;
 P        BoolWrapper url_s0_w = new BoolWrapper();
 P        url_s0_w.b = this.addr_s0_t;
 P        post(url, bytes, url_s0_w, bytes_s1_w);
31      }
        ...
44      public void post(String addr, byte[] bytes,
            BoolWrapper addr_s0_w, BoolWrapper bytes_s1_w){
 P        boolean output_s0_t = addr_s0_w.b;
 P        boolean bytes_s1_t = bytes_s1_w.b;
        ...
48        OutputStream output = conn.getOutputStream();
 P        if(output_s0_t == true && bytes_s1_t == true)
 P          promptForUserDecision();
49        output.write(bytes, 0, bytes.length);
50        ...
51      }
52  }
```

Fig. 4.5 Java code for the patched running example

should also be properly cleaned up. For instance, "addr_s0_t" is set to 0 after Ln.15, where "addr" is assigned a constant value, which means that if no "url" is provided in the Intent, the "addr" should not be tainted.

In the method `onDestroy()`, when the information flows through the `post()` method, we wrap the local shadow variables for corresponding parameters and pass these BoolWrapper objects to new `post()` as additional parameters. In the end, we retrieve the taints in `post()` and check the taint statuses before the critical networking operation is conducted at Ln.49. Consequently, we stop this component hijacking attack right before the dangerous operation takes place.

4.6 Experimental Evaluation

To evaluate the efficacy, correctness and efficiency of AppSealer, we conducted experiments on real-world Android applications with component hijacking vulnerabilities and generated patches for them.

4.6.1 Experiment Setup

We collect 16 vulnerable Android apps, which expose internal capabilities to public interfaces and are subject to exploitation. Table 4.1 describes their exposed interfaces, leaked capabilities and possible security threats.

Table 4.1 Overview of vulnerable apps

ID	Package-version	Exposed interface	Leaked capability	Threat description
1	CN.MyPrivateMessages-52	Activity	Raw query	SQL injection
2	com.akbur.mathsworkout-92	Activity	Internet	Delegation attack
3	com.androidfu.torrents-26	Activity	Selection query	SQL injection
4	com.appspot.swisscodemonkeys.paintfx-4	Activity	Internet	Delegation attack
5	com.cnbc.client-1208	Activity	Selection query	SQL injection
6	com.cnbc.client-1209	Activity	Selection query	SQL injection
7	com.espn.score_center-141	Activity	Internet	Delegation attack
8	com.espn.score_center-142	Activity	Internet	Delegation attack
9	com.gmail.traveldevel.android.vlc.app-131	Service	Internet	Delegation attack
10	com.kmshack.BusanBus-30	Activity	Raw query	SQL injection
11	com.utagoe.momentdiary-45	Service	Raw query	SQL injection
12	com.yoondesign.colorSticker-8	Activity	Raw query	SQL injection
13	fr.pb.tvflash-9	Activity	Selection query	SQL injection
14	gov.nasa-5	Activity	Selection query	SQL injection
15	hu.tagsoft.ttorrent.lite-15	Service	Internet	Delegation attack
16	jp.hotpepper.android.beauty.hair-12	Activity	Raw query	SQL injection

Most of these vulnerable apps accidentally leave their internal `Activities` open and unguarded. Thus, any `Intent` whose target matches the vulnerable one can launch it. Others carelessly accept any `Intent` data from a public `Service` without input validations. Unauthorized external `Intent` can therefore penetrate the app, through these public interfaces, and exploit its internal capabilities. Such leaked capabilities, including SQLite database query and Internet access, are subject to various security threats. For instance, `Intent` data received at the exposed interface may cause SQL Injection; external Intent data sending to Internet may cause delegation attack.

To detect and mitigate component hijacking vulnerabilities, AppSealer automatically performs analysis and rewriting, and generates patched apps. We conduct the experiment on our test machine, which is equipped with Intel(R) Core(TM) i7 CPU (8M Cache, 2.80 GHz) and 8 GB of physical memory. The operating system is Ubuntu 11.10 (64bit).

To verify the effectiveness and evaluate runtime performance of our generated patches, we further run them on a real device. Experiments are carried out on Google Nexus S, with Android OS version 4.0.4.

4.6.2 Summarized Results

We configure AppSealer to take incoming `Intents` from exposed interfaces as sources, and treat outgoing Internet traffic and internal database access as sinks. A taint slice is then a potential path from the `Intent` receiver to these privileged APIs. We compute the slice for each single vulnerable instance, and conduct a quantitative study on it.

Figure 4.6 shows the proportional size of the slices, compared to the total size of the application. We can see that most of the taint slices represent a small portion of entire applications, with the average proportion being 11.7 %. However, we do observe that for a few applications, the slices take up to 45 % of the total size. Some samples (e.g., com.kmshack.BusanBus) are fairly small. Although the total slice size is only up to several thousands Jimple statements, the relative percentage becomes high. Apps like com.cnbc.client operate on incoming `Intent` data in an excessive way, and due to the conservative nature of static analysis, many static and instance fields are involved in the slices.

We also measure the program size on different stages of patch generation and optimizations. We observe that the increase of the program size is roughly proportional to the slice size, which is expected. After patch statement placement, the increased size is, on average, 41.6 % compared to the original program.

Figure 4.7 further visualizes the impact of the four optimizations to demonstrate their performance. The five curves on the figure represent the relative sizes of the program, compared to the original app size, on different processing stages, respectively. The top curve is the program size when patch statement placement has been conducted, while the bottom one stands for the app size after all four patch

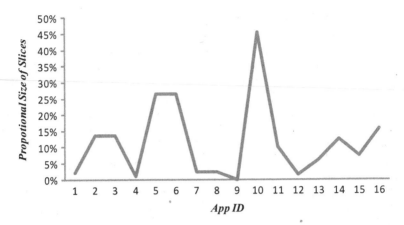

Fig. 4.6 Relative size of slices in percentage

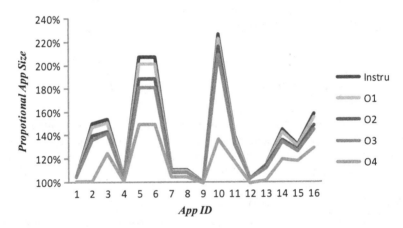

Fig. 4.7 Relative app size in percentage. Five curves quantify app sizes at different stages

optimizations. We can see that (1) for some of these apps, the increase of program size due to patch statement placement can be big, and up to 130 %; and (2) these optimizations are effective, because they significantly reduce the program sizes. In the end, the average size of patch code drops to 15.9 %.

4.6.3   Detailed Analysis

Here we present detail analysis for these vulnerable apps to discuss the effectiveness and accuracy of our generated patches.

4.6.3.1 Apps with Simple Exploiting Paths

Some of the apps are vulnerable but exploitation paths are fairly simple. App 4, 7, 8, 9, 11, 12 fall into this category. Upon obtaining Intent data from an open Activity or Service, these apps directly use it either as URL to load a webpage in WebView, conduct a HTTP GET, or as a SQL string to query an internal database. Consequently, the exploitation path is guaranteed to happen every time a malicious Intent reaches the vulnerable interfaces. In this case, simply blocking the exposed interface might be as good as our patching approach.

However, for other apps, a manipulated input may not always cause an actual exploitation.

4.6.3.2 Apps with Pop-Up Dialogs

Some apps ask user for consent before the capable sink API is called. App 2, 3, 5, 6, 10, 13 share this same feature. If the user does not approve further operations, the exploitation will not occur. In this case, blocking at the open interface causes unnecessary interventions. Our approach, on the other hand, disables the vulnerability and requires only necessary mediations.

`com.akbur.mathsworkout` (version code 92) is one of these examples. This app is a puzzle game, which is subject to data pollution and leaks Internet capability. Granted with Internet permission, the app is supposed to send user's "High Score" to a specific URL. However, the Activity to receive "High Score" data is left unguarded. Thus, an malicious app can send a manipulated Intent with a forged "score" to this vulnerable Activity, polluting the latter's instance field. This field is accessed in another thread and the resulting data is sent to Internet, once the thread is started. However, starting this background thread involves human interactions. Unless a "OK" button is clicked in GUI dialog, no exploitation will happen. Our patch correctly addresses this case and only displays the warning when sending thread is about to call the sink API (i.e. `HttpClient.execute()`).

`com.cnbc.client` (version code 1208 and 1209) asks for user's consent in a more straight-forward way. This finance app exposes an Activity interface that can access internal database, and thus is vulnerable to SQL injection attack. The exposed Activity is intended to receive the "name" of a stock, and further save it to or delete it from the "watch list". Malicious Intent can manipulate this "name" and trick the victim app to delete an important one or add an unwanted one. Nevertheless, the deletion requires user's approval. Before deletion, the app explicitly informs the user and asks for decision. Similarly, if the user chooses "Cancel", no harm will be done. Our patch automatically enforces necessary checks but avoids intervention in this scenario. Notice that taint slices of this app take a great portion (27 %) of the program, and therefore it is extremely hard to confirm and fix the vulnerability, or further discover aforementioned secure path with pure human effort. In contrast, AppSealer automatically differentiates secure and dangerous paths, and in the meantime manages to significantly reduce the amount of patch statements.

4.6.3.3 Apps with Selection Views

A similar but more generic case is that apps provide views such as `AdapterView` for selection. The actual exploit only occurs if an item is selected. Apps 1, 14, 16 are of this kind.

`CN.MyPrivateMessages` (version code 52) is a communication app which suffers the SQL injection attack. An vulnerable Activity may save a manipulated Intent data to its instance field during creation. The app then displays an Adapter-View for user to select call logs. Only upon selection does the event handler obtain data from the polluted field and use it to perform a "delete" operation in database.

4.6.3.4 Apps with Multiple Threads

Some samples extensively create new threads during runtime and pass the manipulated input across threads (e.g., app 2, 10, 15). Asynchronous program execution makes it hard for developers or security analysts to reproduce the exploitation and thus to confirm the vulnerability.

References

1. Davi L, Dmitrienko A, Sadeghi AR, Winandy M (2011) Privilege escalation attacks on Android. In: Proceedings of the 13th international conference on Information security, (Berlin, Heidelberg), 2011
2. Grace M, Zhou Y, Wang Z, Jiang X (2012) Systematic detection of capability leaks in stock Android smartphones. In: Proceedings of the 19th network and distributed system security symposium, 2012
3. Felt AP, Wang HJ, Moshchuk A, Hanna S, Chin E (2011) Permission re-delegation: attacks and defenses. In: Proceedings of the 20th USENIX security symposium, 2011
4. Zhou Y, Jiang X (2013) Detecting passive content leaks and pollution in Android applications. In: Proceedings of the 20th network and distributed system security symposium, 2013
5. Lu L, Li Z, Wu Z, Lee W, Jiang G (2012) CHEX: statically vetting Android apps for component hijacking vulnerabilities. In: Proceedings of the 2012 ACM conference on computer and communications security (CCS'12), October 2012
6. Cui W, Peinado M, Wang HJ (2007) Shieldgen: automatic data patch generation for unknown vulnerabilities with informed probing. In: Proceedings of 2007 IEEE symposium on security and privacy, 2007
7. Brumley D, Newsome J, Song D, Wang H, Jha S (2006) Towards automatic generation of vulnerability-based signatures. In: Proceedings of the 2006 IEEE symposium on security and privacy (Oakland'06), May 2006
8. Costa M, Crowcroft J, Castro M, Rowstron A, Zhou L, Zhang L, Barham P (2005) Vigilante: end-to-end containment of internet worms. In: Proceedings of the twentieth ACM symposium on systems and operating systems principles (SOSP'05), October 2005
9. Costa M, Castro M, Zhou L, Zhang L, Peinado M (2007) Bouncer: securing software by blocking bad input. In: Proceedings of 21st ACM SIGOPS symposium on operating systems principles (SOSP'07), October 2007

10. Caballero J, Liang Z, Poosankam, Song D (2009) Towards generating high coverage vulnerability-based signatures with protocol-level constraint-guided exploration. In: Proceedings of the 12th international symposium on recent advances in intrusion detection (RAID'09), September 2009
11. Lin Z, Jiang X, Xu D, Mao B, Xie L (2007) AutoPAG: towards automated software patch generation with source code root cause identification and repair. In: Proceedings of the 2nd ACM symposium on information, computer and communications security, 2007
12. Zhang C, Wang T, Wei T, Chen Y, Zou W (2010) IntPatch: automatically fix integer-overflow-to-buffer-overflow vulnerability at compile-time. In: Proceedings of the 15th European conference on research in computer security, 2010.
13. Sidiroglou S, Keromytis AD (2005) Countering network worms through automatic patch generation. In: IEEE security and privacy, Nov 2005, vol 3, pp 41–49
14. Newsome J (2006) Vulnerability-specific execution filtering for exploit prevention on commodity software. In: Proceedings of the 13th symposium on network and distributed system security (NDSS), 2006
15. Soot: A Java Optimization Framework (2016) http://www.sable.mcgill.ca/soot/
16. dex2jar (2016) http://code.google.com/p/dex2jar/

Chapter 5
Efficient and Context-Aware Privacy Leakage Confinement

Abstract As Android has become the most prevalent operating system in mobile devices, privacy concerns in the Android platform are increasing. A mechanism for efficient runtime enforcement of information-flow security policies in Android apps is desirable to confine privacy leakage. The prior works towards this problem require firmware modification (i.e., modding) and incur considerable runtime overhead. Besides, no effective mechanism is in place to distinguish malicious privacy leakage from those of legitimate uses. In this paper, we take a bytecode rewriting approach. Given an unknown Android app, we selectively insert instrumentation code into the app to keep track of private information and detect leakage at runtime. To distinguish legitimate and malicious leaks, we model the user's decisions with a context-aware policy enforcement mechanism. We have implemented a prototype called *Capper* and evaluated its efficacy on confining privacy-breaching apps. Our evaluation on 4723 real-world Android applications demonstrates that Capper can effectively track and mitigate privacy leaks. Moreover, after going through a series of optimizations, the instrumentation code only represents a small portion (4.48 % on average) of the entire program. The runtime overhead introduced by Capper is also minimal, merely 1.5 % for intensive data propagation.

5.1 Introduction

Privacy concerns in the Android platform are increasing. Previous studies [1–6] have exposed that both benign and malicious apps are stealthily leaking users' private information to remote servers. Efforts have also been made to detect and analyze privacy leakage either statically or dynamically [1, 2, 7–11]. Nevertheless, a good solution to defeat privacy leakage at runtime is still lacking. We argue that a practical solution needs to achieve the following goals:

- **Information-flow based security.** Privacy leakage is fundamentally an information flow security problem. A desirable solution to defeat privacy leakage would detect sensitive information flow and block it right at the sinks. However, most of prior efforts to this problem are "end-point" solutions. Some earlier solutions extended Android's install-time constraints and enriched Android permissions [12, 13]. Some aimed at enforcing permissions in a finer-grained

M. Zhang, H. Yin, *Android Application Security*, SpringerBriefs in Computer Science, DOI 10.1007/978-3-319-47812-8_5

63

manner and in a more flexible way [14–17]. Some attempted to improve isolation on various levels and each isolated component could be assigned with a different set of permissions [18–20]. In addition, efforts were made to introduce supplementary user consent acquisition mechanism, so that access to sensitive resource also requires user approval [21, 22]. All these "end-point" solutions only mediate the access to private information, without directly tackling the privacy leakage problem.

- **Low runtime overhead.** An information-flow based solution must have very low runtime overhead to be adopted on end users' devices. To directly address privacy leakage problem, Hornyack et al. proposed AppFence to enforce information flow policies at runtime [3]. With support of TaintDroid [2], AppFence keeps track of the propagation of private information. Once privacy leakage is detected, AppFence either blocks the leakage at the sink or shuffle the information from the source. Though effective in terms of blocking privacy leakage, its efficiency is not favorable. Due to the taint tracking on every single Dalvik bytecode instruction, AppFence incurs significant performance overhead.
- **No firmware modding.** For a practical solution to be widely adopted, it is also crucial to avoid firmware modding. Unfortunately, the existing information-flow based solutions such as AppFence require modifications on the stock software stack, making it difficult to be deployed on millions of mobile devices.
- **Context-aware policy enforcement.** Many apps need to access user's privacy for legitimate functionalities and these information flows should not be stopped. Therefore, to defeat privacy leakage without compromising legitimate functionality, a good solution needs to be aware of the context where a sensitive information flow is observed and make appropriate security decisions. To the best of our knowledge, we are not aware that such a policy mechanism exists.

In this chapter, we aim to achieve all these design goals by taking a bytecode rewriting approach. Given an unknown Android app, we selectively rewrite the program by inserting bytecode instructions for tracking sensitive information flows *only* in certain fractions of the program (which are called taint slices) that are potentially involved in information leakage. When an information leakage is actually observed at a sink node (e.g., an HTTP Post operation), this behavior along with the program context is sent to the policy management service installed on the device and the user will be notified to make an appropriate decision. For example, the rewritten app may detect the location information being sent out to a Google server while the user is navigating with Google Map, and notify the user. Since the user is actively interacting with the device and understands the context very well, he or she can make a proper decision. In this case, the user will allow this behavior. To ensure good user experiences, the number of such prompts must be minimized. To do so, our policy service needs to accurately model the context for the user's decisions. As a result, when an information leakage happens in the same context, the same decision can be made without raising a prompt. After exploring the design space of the context modeling and making a balance between sensitivity, performance overhead, and robustness, we choose to model the context using *parameterized source and sink pairs*.

Consequently, our approach fulfills all the requirements: (1) actual privacy leaks are captured accurately at runtime, with the support of inserted taint tracking code; (2) the performance overhead of our approach is minimal, due to the static dataflow analysis in advance and numerous optimizations that are applied to the instrumentation code; (3) the deployment of our approach is simple, as we only rewrite the original app to enforce certain information flow policies and no firmware modification is needed; (4) policy enforcement is context-aware, because the user's decisions are associated with abstract program contexts.

We implement a prototype, *Capper*,[1] in 16 thousand lines of Java code, based on the Java bytecode optimization framework Soot [23]. We leverage Soot's capability to perform static dataflow analysis and bytecode instrumentation. We evaluate our tool on 4723 real-world privacy-breaching Android apps. Our experiments show that rewritten programs run correctly after instrumentation, while privacy leakage is effectively eliminated.

5.2   Approach Overview

5.2.1   Key Techniques

Figure 5.1 depicts an overview of our techniques. When a user is about to install an app onto his Android device, this app will go through our bytecode rewriting engine (BRIFT) and be rewritten into a new app, in which sensitive information flows are monitored by the inserted bytecode instructions. Therefore, when this

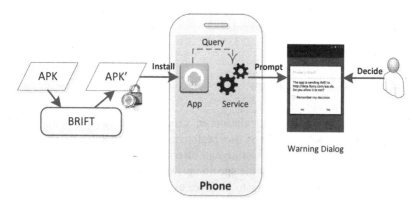

Fig. 5.1 Architecture of Capper

---

[1]*Capper* is short for Context-Aware Privacy Policy Enforcement with Re-writing.

new app is actually running on the device and is observed to send out sensitive information, this behavior (along with the program context) will be reported to the policy management service for decision.

If this behavior under this program context is observed for the first time, the policy management service will prompt the user to make a proper decision: either allow or deny such a behavior. The user's decision will be recorded along with the program context, so the policy management service will make the recorded decision for the same behaviors under the same context.

Therefore, our solution to defeat privacy leakage consists of the following two enabling techniques.

(1) **Bytecode Rewriting for Information Flow Control.** Given a bytecode program, the goal of our bytecode rewriting is to insert a minimum amount of bytecode instructions into the bytecode program to trace the propagation of certain sensitive information flows (or taint). To achieve this goal, we first conduct static dataflow analysis to compute a number of program slices that are involved in the taint propagation. Then we insert bytecode instructions along the program slices to keep track of taint propagation at runtime. Further, we perform a series of optimizations to reduce the amount of inserted instructions. Please refer to Chap. 4 for more details.

(2) **Context-aware Policy Enforcement.** The user allows or denies a certain information flow in a specific context. The key for a context-aware policy enforcement is to properly model the context. The context modeling must be sensitive enough to distinguish different program contexts, but not too sensitive. Otherwise, a slight difference in the program execution may be treated as a new context and may cause unnecessarily annoying prompts to the user. Further, the context modeling should also be robust enough to counter mimicry attacks. An attacker may be able to "mimic" a legitimate program context to bypass the context-aware policy enforcement.

5.3   Context-Aware Policy

Once our inserted monitoring code detects an actual privacy leakage, policy service will enforce privacy policy based on user preferences. To be specific, the service app inquires user's decision upon detection and offers the user options to either "one-time" or "always" allow or deny the specific privacy breaching flow. The user can then make her decision according to her user experience and the policy manager will remember users preference for future decision making situations if the "always" option is chosen.

There exist two advantages to enforce a privacy policy with user preference history. Firstly, it associates user decisions with certain program contexts and can thus selectively restrict privacy-related information flow under different circumstances. Privacy-related outbound traffic occurs in both benign and malicious semantics.

However, from dataflow analysis perspective, it is fairly hard to distinguish between, for example, a coordinates-based query towards a benign map service and a location leakage via some covert malicious process within, say, a wallpaper app. On the contrary, it is fairly straight-forward for a user to tell the difference because she knows the application semantics. With human knowledge, it is possible to avert overly strict restriction and preserve usability to the largest extent.

Secondly, it avoids repetitive warning dialogs and improves user experience. Once an "always" decision is made, this decision will be remembered and used for the same scenario next time. Thus, the user doesn't need to face the annoying dialog message over and over again for the exactly identical situation.

However, it is non-trivial to appropriately model the program context specific to a user decision and the challenge lies in the way semantics is extracted from a dataflow point of view. We hereby discuss some possible options and our solution.

5.3.1   Taint Propagation Trace

To achieve high accuracy, we first consider using the exact execution trace as pattern to represent a specific information flow. An execution trace can be obtained at either instruction or method level. It consists of all the instructions or methods propagating sensitive data from a source to a sink, and therefore can uniquely describe a dataflow path. Formally, we define a trace-based context as $C_T = [t_0, t_1, \ldots, t_n]$, where each t_i represents an individual instruction or method. When a user decision is made for a certain information flow, its execution trace is computed and saved as a pattern along with user preference. Next time when a new leakage instance is detected, the trace computation will be done on the new flow and compared with saved ones. If there exists a match, action taken on the saved one will be applied correspondingly.

Nevertheless, there exist two major drawbacks with this approach. Firstly, dynamic tracing is considerably heavy-weight. Comparison of two traces is also fairly expensive. This may affect the responsiveness of interactive mobile apps. Secondly, each dataflow instance is modeled overly precisely. Since any execution divergence will lead to a different trace pattern, even if two leakage flows occur within the same semantics, it is still difficult to match their traces. This results in repeated warning messages for semantically equivalent privacy-related dataflows.

5.3.2   Source and Sink Call-Sites

Trace-based approach is too expensive because the control granularity is extremely fine. We therefore attempt to relax the strictness and achieve balance in the accuracy-efficiency trade-off. We propose a call-site approach which combines source-sink call-sites to model privacy flow. That is to say, information flows of same source and sink call-sites are put into one category. Thus, such a context is formally defined to be $C_{SS} = \{Source, Sink\}$. Once an action is taken on one leakage flow, the same action will be taken on future sensitive information flow in the same category. To this end, we introduce labels for source and sink call-sites. Information flows starting from or arriving at these call-sites are associated with corresponding labels, so that they can be differentiated based on these labels.

With a significant improvement of efficiency, this approach is not as sensitive to program contexts as the traced-based one—different execution paths can start from the same origin and end at the same sink. However, we rarely observed this inaccuracy in practice because the app execution with same source and sink call-sites usually represents constant semantics.

5.3.3   Parameterized Source and Sink Pairs

In addition to source/sink call-sites, parameters fed into these call-sites APIs are also crucial to the semantic contexts. For example, the user may allow an app to send data to certain trustworthy URLs but may not be willing to allow access to the others. Therefore, it is important to compare critical parameters to determine if a new observed flow matches the ones in history. We therefore define this parameterized callsites based context as a triple $C_{PSS} = \{Source, Sink, Params\}$, where $Params$ contains the critical parameters that are consumed by the two callsites.

Notice that checking parameters can minimize the impact of mimicry attack. Prior research shows that vulnerable Android apps are subject to various attacks [7, 24–27]. For instance, an exposed vulnerable app component can be exploited to leak private information to an attacker-specified URL. Without considering the URL parameter, it is difficult, if not possible, to distinguish internal use of critical call-sites from hijacking the same call-sites to target a malicious URL. Once a flow through some call-sites is allowed and user preference is saved, mimicking attack using same call-sites will also get approved. On the contrary, a parameter-aware approach can differentiate outgoing dataflows according to the destination URL, and thus, exploitation of a previously allowed call-sites will still raise a warning.

Besides the URL of a Internet API, we also consider some other critical combinations of an API and its parameter. Table 5.1 summarizes our list. The target of a sink API is sensitive to security. Similar to Internet APIs, target phone numbers are crucial to `sendTextMessage()` APIs and thus need watching. On the other hand, some source call-sites also need to be distinguished according to the

Table 5.1 APIs and critical parameters

API description	Source or sink	Critical parameter
Send data to internet	Sink	Destination URL
Send SMS message	Sink	Target phone number
Query contacts database	Source	Source URI

parameters. For instance, the API that queries contacts list may obtain different data depending on the input URI (e.g., `ContactsContract.CommonDataKinds.Phone` for phone number, `ContactsContract.CommonDataKinds.Email` for email).

5.3.4 Implementation

We implement the policy service as a separate app. This isolation guarantees the security of the service app and its saved policies. In other words, even if the client is exploited, the service is not affected or compromised, and can still correctly enforce privacy policies.

The service app communicates with a rewritten client app solely through Android IPC. Once a client app wants to query the service, it encapsulates the labels of source and sink call-sites as well as the specific critical parameter into an `Intent` as extra data payload. The client app is then blocked and waiting for a response. Since this transaction is usually fast, the blocking won't affect the responsiveness of the app mostly. On the service side, it decodes the data and searches for a match in its database. If there exists a match, it returns immediately with the saved action to the client. Otherwise, the service app will display a dialog message within a created `Activity`. The user decision is saved if the user prefers, or not saved otherwise. Either way, user's option is sent back to the client. On receiving the response from service, the rewritten app will either continue its execution or skip the sink call-site with respect to the reply.

It is noteworthy that we have to defend against spoofing attack and prevent forged messages from being sent to a client app. To address that, we instrument the client app to listen for the service reply with a dynamically registered `BroadcastReceiver`. When a broadcast message is received, the receiver is immediately unregistered. Thus, attack window is reduced due to this on-demand receiver registration. Further, to restrict who can send the broadcast, we protect the receiver with a custom `permission`. Broadcaster without this `permission` is therefore unable to send messages to the client app. To defeat replay attack, we also embed a session token in the initial query message, and the client app can therefore authenticate the sender of a response message.

Similarly, we need to protect a policy service from spoofing, too. The service app has to check the caller identity from a bound communication via `getCallingUid()`, so that a malicious application cannot pretend to be another app and trick the service to configure the policies for the latter.

5.4 Experimental Evaluation

To evaluate the efficacy, correctness and efficiency of Capper, we conducted experiments on real-world Android applications. In the policy setting, we consider IMEI, owner's phone number, location, contacts to be the sensitive information sources, and network output APIs (e.g., `OutputStream.write()`, `HttpClient.execute()`, `WebView.loadUrl()`, `URLConnection.openConnection()` and `SmsManager.sendTextMessage()`) as the sinks. The action on the sink can be "block" or "allow". User can also check an "always" option to have the same rule applied to future cases of the same semantic context. Note that this policy is mainly for demonstrating the usability of Capper. More work is needed to define a more complete policy for privacy leakage confinement.

We obtain 4915 applications from Google Play and use them as our experiment sample set. We then perform bytecode rewriting on these apps to enable runtime privacy protection. We conduct the experiment on our test machine, which is equipped with Intel(R) Xeon(R) CPU E5-2690 (20 M Cache, 2.90 GHz) and 200 GB of physical memory. The operating system is CentOS 6.3 (64 bit). To verify the effectiveness and evaluate runtime performance of the rewritten apps, we further run them on a real device. Experiments are carried out on Google Nexus S, with Android OS version 4.0.4.

5.4.1 Summarized Analysis Results

Figure 5.2 illustrates the partition of 4915 realworld apps. Amongst these apps, Capper did not finish analyzing 314 of them within 30 min. These apps are fairly large (many over 10 MB). Application-wide dataflow analysis is known to be expensive for these large programs. We further extended the analysis timeout to 3 h, and 122 more apps were successfully analyzed and rewritten. Given sufficient analysis time, we believe that the success rate can further increase from currently 96 % to nearly 100 %.

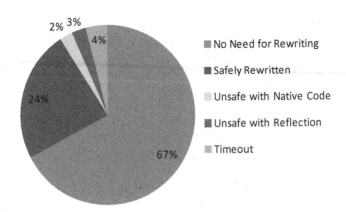

Fig. 5.2 Bytecode rewriting results on 4915 realworld android apps

Out of the 4723 apps that were completely processed by Capper, 1414 apps may leak private information, according to our static analysis, and Capper successfully performed bytecode rewriting on them. We observed that most of them leak IMEI and location information. These types of information are frequently sent out by apps due to analytic or advertisement reasons. Apps may also sometimes leak owner's phone number or phone numbers from contacts. For the rest of them (67 %), static analysis couldn't find a viable path from the sensitive information sources to the network output sinks. It means that these apps do not leak private information, so no rewriting is needed for these apps.

For these 1414 apps that were rewritten, we further investigate their use of native code and reflection. Our study shows that unknown native code is invoked within the taint slices for 118 (2 %) apps. As a bytecode-level solution, our system cannot keep track of information flow processing in the native code. So the information flow may be broken on these unknown native calls. The rest 3 % contain reflective calls within the slices. If the class name and method name cannot be resolved statically, we do not know how information is propagated through this function. Therefore, totally 5 % apps may be unsafe and may not be fully enforced with the specified policies. The best suggestion for the end user is not to use these unsafe rewritten apps due to the potential incompleteness in policy enforcement.

We compute the taint propagation slice for each single leakage instance, and conduct a quantitative study on them. While most of the apps retrieve privacy-related information moderately, some apps leak user's privacy through up to 31 taint slices. Such apps usually enclose various Ads libraries, each of which acquires private data separately.

Most of the program slices represent a small portion of the entire application, with the average proportion being 2.48 %. However, we do observe that for a few applications, the slices take up to 54 % of the total size. Some samples (e.g., com.tarsin.android.dilbert) are fairly small. Although the total slice size is only up to several thousands Jimple statements, the relative percentage becomes high. Apps like de.joergjahnke.mario.android.free and de.schildbach.oeffi operate on privacy information in an excessive way, and due to the conservative nature of static analysis, many static and instance fields are involved in the slices.

We measure the program size on different stages. We observe that the increase of the program size is roughly proportional to the slice size, which is as expected. After instrumentation, the increased size is, on average, 10.45 % compared to the original program. The proposed optimizations are effective and they significantly reduce the program sizes. In the end, the average size of inserted code drops to 4.48 %.

5.4.2 Detailed Analysis

Here we present the detailed analysis results of ten applications, and demonstrate the effectiveness of context-aware privacy policy enforcement. To this end, we rewrite these apps with Capper, and run them in a physical device along with our policy service app. We further manually trigger the privacy leakage components in the app, so that inserted code would block the program and query policy service for decision. The service will then search its own database to see if there exists a rule for the specific dataflow context of the requesting app. If a corresponding rule exists, service replies to requester immediately. Otherwise, a dialog is displayed to the user asking for decision. The user can also check the "always" option so that current decision will be saved for further reference (Fig. 5.3). Notice that, in order to test the context-awareness of our approach, we always check this option during the experiment. Therefore, from the logcat information on the service side, we may observe and compare the number of queries an app makes with the amount of warning dialogs prompted to the user. We also compute the number of information flow contexts with trace-based model for comparison.

Table 5.2 lists the summarized results including number of queries, prompts and trace-based contexts. For these apps, the prompt number is often equal to

Fig. 5.3 Warning dialog

Table 5.2 Effectiveness of context-aware policy enforcement

ID	App-Version	Queries	Prompts	Trace-based contexts
1	artfulbits.aiMinesweeper-3.2.3	5	2	2
2	com.avantar.wny-4.1.5.1	3	2	2
3	com.avantar.yp-4.1.5	4	2	2
4	com.bfs.papertoss-1.09	3	2	2
5	com.rs.autorun-1.3.1	10	4	6
6	com.skyfire.browser-2.0.1	4	2	2
7	com.startapp.wallpaper.browser-1.4.15	5	3	3
8	mabilo.ringtones.app-3.6	5	3	3
9	mabilo.wallpapers-1.8.4	4	2	2
10	net.flixster.android-2.9.5	6	3	3

or sometimes slightly smaller than the amount of trace-based dataflow contexts, while the number of queries is usually much larger than that of prompts. This means leakage contexts are modeled correctly: disparate contexts are usually treated differently in our callsite-based approach; and equivalent contexts are enforced with the same rule. The fundamental reason is that Android apps are often componentized while a separate component exercises a dedicated function.

Firstly, different types of private information are accessed through separate execution paths. Some apps (ID 2, 3) retrieve both device identifier and location information, and send them at separate sinks. Similarly, mabilo.ringtones.app leaks both geolocation data and user's phone number.

Secondly, the same type of privacy-related data can be retrieved from isolated packages but serves as different purposes. These apps (ID 4, 7, 10) consist of a main program and advertisement components, both of which produce outgoing traffic taking IMEI or location data. Take net.flixster.android as an example. In this movie app, location data is both sent to flixster server for querying theaters and to Ads server for analytical purpose.

Further, the same private data can be accessed in one package but contributes to different use cases. For instance, com.rs.autorun obtains device identifier via getDeviceId() API call in the first place. Then the app sends IMEI at multiple sinks to load advertisement View, retrieve configuration or upload analytical information including demographics, ratings, etc. Each semantic context is captured from the perspective of taint propagation trace. However, due to the use of same sink call-site, all the analytical traffics are considered to be with the same semantic context in the call-site model. Though call-site-based model is not as sensitive to context as the trace-based approach, it is still able to differentiate the contexts of Ads View, configuration file loading and analytical dataflow, according to disparate sink call-sites.

We also observe inconsistency between traced-based contexts and real program semantics. That lies in apps (ID 1, 6, 7, 8, 9, 10) which acquire location information, where the number of trace-based contexts exceeds that of actual contexts. Geographic location is obtained either approximately from getLastKnownLocation(), or from a real-time update by registering a listener through requestLocationUpdates() and receiving updates via callback onLocationChanged(Location). Some apps adopt both ways so as to achieve higher accuracy. For instance, artfulbits.aiMinesweeper reads location data by calling getLastKnownLocation() at the very beginning of the program, stores it into an instance field, and then periodically updates it with the aforementioned callback. Consequently, two separate paths achieve one sole purpose and thus should be considered as of equivalent context. However, from either trace or call-site point of view, there exist two separate contexts. Despite the disparity, we believe that this would at most introduce one extra prompt. Further, it is also reasonable to split this context into two, because one conducts a one-time access while the other obtains data repeatedly.

5.4.3 Runtime Performance

We compare the runtime overhead of bytecode rewriting with that of dynamic taint analysis in TaintDroid (on Android gingerbread version 2.3.4). In principle, if an app leaks private information only occasionally, the rewritten version would have much better performance than the original version on TaintDroid. This is because in the rewritten app nearly no instrumentation code is added on non-leaking execution paths whereas TaintDroid has to monitor taint propagation all the time.

Rather, we would like to compare the performance when the taint is actively propagated during the execution. This would be the worst-case scenario for Capper. Specifically, we build two customized applications for the measurement. Both leak IEMI string via `getDeviceId()` API, decode the string to a byte array, encrypt the array by doing XOR on every byte with a constant key, reassemble a string from the encrypted byte array, and in the end, send the string out to a specific URL through a raw socket interface. The only difference is that one merely sends the IMEI string, while the other also appends extra information of totally 10 KB to the IMEI string before encryption. In other words, the former conducts a short-period data transfer while the latter manipulates the outgoing message within a much longer period. We expect that the execution of the first one to be mainly under `mterp` interpretation mode and the execution of the second to be boosted by JIT.

We measured the execution time from `getDeviceId()` to the network interface. We observed that the rewritten application runs significantly faster than the original on TaintDroid, and only yields fairly small overhead compared to the original one running on Android, for both short-period and long-period data propagation. Table 5.3 illustrates the result of runtime measurement. While our approach causes 13 and 1.5 % overhead for short and long data propagation respectively, TaintDroid incurs 330 and 47 % overhead. The results also show that the presence of JIT significantly reduces runtime overhead, in both approaches. However, though newer version of TaintDroid (2.3.4 and later) benefits from JIT support, the overhead caused by dynamic instrumentation is still apparently high.

To further confirm the runtime overhead of the rewritten programs, we conduct an experiment on Google Nexus S, with Android OS version 4.0.4. It is worth mentioning that such verification on real device requires considerable repetitive manual efforts and thus is fairly time consuming. We therefore randomly pick 10 apps from the privacy-breaching ones, rewrite them, run both the original app and secured one on physical device, and compare the runtime performance before and after rewriting. We rely on the timestamps of Android framework debugging information (logcat logs) to compute the app load time as benchmark. The app load time is measured from when Android `ActivityManager` starts an `Activity`

Table 5.3 Runtime performance evaluation

	Orig.	Orig. on TaintDroid	Rewritten
Short	30 ms	130 ms	34 ms
Long	10,583 ms	15,571 ms	10,742 ms

component to the time the `Activity` thread is displayed. This includes application resolution by `ActivityManager`, IPC and graphical display. Our observation complies with prior experiment result: rewritten apps usually have insignificant slowdown, with an average of 2.1 %, while the maximum runtime overhead is less than 9.4 %.

References

1. Enck W, Octeau D, McDaniel P, Chaudhuri S (2011) A study of Android application security. In: Proceedings of the 20th usenix security symposium, August 2011
2. Enck W, Gilbert P, Chun BG, Cox LP, Jung J, McDaniel P, Sheth AN (2010) TaintDroid: an information-flow tracking system for realtime privacy monitoring on smartphones. In: Proceedings of the 9th USENIX symposium on operating systems design and implementation (OSDI'10), October 2010
3. Hornyack P, Han S, Jung J, Schechter S, Wetherall D (2011) These aren't the droids you're looking for: retrofitting Android to protect data from imperious applications. In: Proceedings of CCS, 2011
4. Zhou Y, Jiang X (2012) Dissecting Android malware: characterization and evolution. In: Proceedings of the 33rd IEEE symposium on security and privacy (Oakland'12), May 2012
5. Zhou Y, Wang Z, Zhou W, Jiang X (2012) Hey, you, get off of my market: detecting malicious apps in official and alternative Android markets. In: Proceedings of 19th annual network and distributed system security symposium (NDSS'12), February 2012
6. Wu C, Zhou Y, Patel K, Liang Z, Jiang X (2014) AirBag: boosting smartphone resistance to malware infection. In: Proceedings of the 21th annual network and distributed system security symposium (NDSS'14), February 2014
7. Lu L, Li Z, Wu Z, Lee W, Jiang G (2012) CHEX: statically vetting Android apps for component hijacking vulnerabilities. In: Proceedings of the 2012 ACM conference on computer and communications security (CCS'12), October 2012
8. Gibler C, Crussell J, Erickson J, Chen H (2012) AndroidLeaks: automatically detecting potential privacy leaks in Android applications on a large scale. In: Proceedings of the 5th international conference on Trust and Trustworthy Computing, 2012
9. Kim J, Yoon Y, Yi K, Shin J (2012) Scandal: Static Analyzer for Detecting Privacy Leaks in Android Applications. In: Mobile Security Technologies (MoST) 2012
10. Mann C, Starostin A (2012) A framework for static detection of privacy leaks in Android applications. In: Proceedings of the 27th annual ACM symposium on applied computing, 2012
11. Yang Z, Yang M, Zhang Y, Gu G, Ning P, Wang XS (2013) AppIntent: analyzing sensitive data transmission in Android for privacy leakage detection. In: Proceedings of the 20th ACM conference on computer and communications security (CCS'13), November 2013
12. Ongtang M, McLaughlin S, Enck W, McDaniel P (2009) Semantically rich application-centric security in Android. In: Proceedings of ACSAC, 2009
13. Enck W, Ongtang M, McDaniel P (2009) On lightweight mobile phone application certification. In: Proceedings of the 16th ACM conference on computer and communications security (CCS'09), November 2009
14. Conti M, Nguyen VTN, Crispo B (2011) Crepe: context-related policy enforcement for Android. In: Proceedings of the 13th international conference on information security, 2011
15. Zhou Y, Zhang X, Jiang X, Freeh VW (2011) Taming information-stealing smartphone applications (on Android). In: Proceedings of the 4th international conference on Trust and trustworthy computing, 2011
16. Nauman M, Khan S, Zhang X (2010) Apex: extending Android permission model and enforcement with user-defined runtime constraints. In: Proceedings of the 5th ACM symposium on information, computer and communications security, 2010

17. Beresford AR, Rice A, Skehin N, Sohan R (2011) Mockdroid: trading privacy for application functionality on smartphones. In: Proceedings of the 12th workshop on mobile computing systems and applications, 2011

18. Lange M, Liebergeld S, Lackorzynski A, Warg A, Peter M (2011) L4Android: a generic operating system framework for secure smartphones. In: Proceedings of the 1st ACM workshop on security and privacy in smartphones and mobile devices, 2011

19. Andrus J, Dall C, Hof AV, Laadan O, Nieh J (2011) Cells: a virtual mobile smartphone architecture. In: Proceedings of SOSP, 2011

20. Shekhar S, Dietz M, Wallach DS (2012) Adsplit: separating smartphone advertising from applications. In: Proceedings of the 20th usenix security symposium, August 2012

21. Xu R, Sadi H, Anderson R (2012) Aurasium: practical policy enforcement for Android applications. In: Proceedings of the 21th usenix security symposium, August 2012

22. Livshits B, Jung J (2013) Automatic mediation of privacy-sensitive resource access in smartphone applications. In: Proceedings of the 22th usenix security symposium, 2013

23. Soot: A Java Optimization Framework (2016) http://www.sable.mcgill.ca/soot/

24. Felt AP, Wang HJ, Moshchuk A, Hanna S, Chin E (2011) Permission re-delegation: attacks and defenses. In: Proceedings of the 20th USENIX security symposium, 2011

25. Grace M, Zhou Y, Wang Z, Jiang X (2012) Systematic detection of capability leaks in stock Android smartphones. In: Proceedings of the 19th network and distributed system security symposium, 2012

26. Zhou Y, Jiang X (2013) Detecting passive content leaks and pollution in Android applications. In: Proceedings of the 20th network and distributed system security symposium, 2013

27. Davi L, Dmitrienko A, Sadeghi AR, Winandy M (2011) Privilege escalation attacks on Android. In: Proceedings of the 13th international conference on information security, Berlin, Heidelberg, 2011

Chapter 6
Automatic Generation of Security-Centric Descriptions for Android Apps

Abstract To improve the security awareness of end users, Android markets directly present two classes of literal app information: (1) permission requests and (2) textual descriptions. Unfortunately, neither can serve the needs. A permission list is not only hard to understand but also inadequate; textual descriptions provided by developers are not security-centric and are significantly deviated from the permissions. To fill in this gap, we propose a novel technique to automatically generate security-centric app descriptions, based on program analysis. We implement a prototype system, DESCRIBEME , and evaluate our system using both DroidBench and real-world Android apps. Experimental results demonstrate that DESCRIBEME enables a promising technique which bridges the gap between descriptions and permissions. A further user study shows that automatically produced descriptions are not only readable but also effectively help users avoid malware and privacy-breaching apps.

6.1 Introduction

Unlike traditional desktop systems, Android provides end users with an opportunity to proactively accept or deny the installation of any app to the system. As a result, it is essential that the users become aware of each app's behaviors so as to make appropriate decisions. To this end, Android markets directly present the consumers with two classes of information regarding each app: (1) the Android permissions requested and (2) textual description of the app's behavior that is provided by the app's developer. Unfortunately, neither of these can fully serve this need.

Permission requests are not easy to understand. First, prior study [1] has shown that few users are cautious or knowledgeable enough to comprehend the security implications of Android permissions. Second, a permission list merely tells the users which permissions are used, but does not explain *how* they are used. Without such knowledge, one cannot properly assess the risk of allowing a permission request. For instance, both a benign navigation app and a spyware instance of the same app can require the same permission to access GPS location, yet use it for completely different purposes. While the benign app delivers GPS data to a legitimate map server upon the user's approval, the spyware instance can periodically and stealthily

© The Author(s) 2016
M. Zhang, H. Yin, *Android Application Security*, SpringerBriefs in Computer
Science, DOI 10.1007/978-3-319-47812-8_6

leak the user's location information to an attacker's site. Due to the lack of context clues, a user is not able to perceive such differences via the simple permission enumeration.

Textual descriptions provided by developers are not security-centric. There exists very little incentive for app developers to describe their products from a security perspective, and it is still a difficult task for average developers (usually inexperienced) to write dependable descriptions. Malware authors can also intentionally hide malice from innocent users by providing misleading descriptions. Previous studies [2, 3] have revealed that the existing descriptive text deviate considerably from requested permissions. As a result, developer-driven description generation cannot be considered trustworthy.

To address this issue, we propose a novel technique to automatically generate app descriptions which accurately describe the security-related behaviors of Android apps. To interpret panoramic app behaviors, we extract *security behavior graphs* as high-level program semantics. To create concise descriptions, we further condense the graphs by mining and compressing the frequent subgraphs. As we traverse and parse these graphs, we leverage *Natural Language Generation* (NLG) to automatically produce concise, human-understandable descriptions.

A series of efforts have been made to describe the functionalities of traditional Java programs as human readable text via NLG. Textual summaries are automatically produced for methods [4], method parameters [5], classes [6], conditional code snippets [7] and algorithmic code structures [8] through program analysis and comprehension. These works focus upon depicting the intra-procedural structure-based operations, while our technique presents the whole-program's semantic-level activities. Furthermore, we take the first step towards automating Android app description generation for security purposes.

We implement a prototype system, DESCRIBEME , in 25 thousand lines of Java code. Our behavior graph generation is built on top of Soot [9], while our description production leverages an NLG engine [10] to realize texts from the graphs. We evaluate our system using both DroidBench [11] and real-world Android apps. Experimental results demonstrate that DESCRIBEME is able to effectively bridge the gap between descriptions and permissions. A further user study shows that our automatically-produced descriptions are both readable and effective at helping users to avoid malware and privacy-breaching apps.

6.2   Overview

6.2.1   Problem Statement

Figures 6.1 and 6.2 demonstrate the two classes of descriptive metadata that are associated with an Android app available via Google Play. The app shown leaks the

Fig. 6.1 Permission requests

user's phone number and service provider to a remote site. Unfortunately, neither of these two pieces of metadata can effectively inform the end users of the risk.

The permission list (Fig. 6.1) simply enumerates all of the permissions requested by the app while replacing permission primitives with straightforward explanations. Besides, it can merely tell users that the app uses two separate permissions, READ_PHONE_STATE and INTERNET, but cannot indicate that these two permissions are used consecutively to send out phone number.

The textual descriptions are not focused on security. As depicted in the example (the top part in Fig. 6.2), developers are more interested in describing the app's functionalities, unique features, special offers, use of contact information, etc. Prior studies [2, 3] have revealed significant inconsistencies between app descriptions and permissions.

We propose a new technique, DESCRIBEME, which addresses these shortcomings and can automatically produce complementary security-centric descriptions for apps in Android markets. It is worth noting that we do not expect to replace the developers' descriptions with our own. Instead, we hope to provide additional app information that is written from a security perspective. For example, as shown in the bottom part of Fig. 6.2, our security-sensitive descriptions are attached to the

Fig. 6.2 Old+New descriptions

× Free international phone calls ...

◆ **International calls are not free** ◆
Free International Calls - Call Sky Features
- Free international calls using premium line
- Excellent call quality
- Join / Register is convenient
◆ **Is it really free?** ◆
Sky calls charged only domestic calls in accordance with the user's smartphone plans, and international telephone charges will not occur at all.
◆ **How to use?** ◆
Including the country code in the dial screen, press the call button after input target number.
◆ **Service Contact** ◆
Home page: ~~www.~~
Contact:~~

Once a GUI component is clicked, the app retrieves your phone number and encodes data in the format "100/app_id=an1005/ani=%s/dest=%s /phone_number=%s/company=%s/" and sends the data to network depending on if the user selects Button "Confirm".

Once a GUI component is clicked, the app retrieves the service provider name and encodes data in the format "100/app_id=an1005/ani=%s /dest=%s/phone_number=%s/company=%s/" and sends the data to network depending on if the user selects Button "Confirm".

existing ones. The new description states that the app retrieves the phone number and writes data to network, and therefore indicates the privacy-breaching behavior.

We expect to primarily deploy DESCRIBEME directly into the Android markets, as illustrated in Fig. 6.3. Upon receiving an app submission from a developer, the market drives our system to analyze the app and create a security-centric description. The generated descriptions are then attached to the corresponding apps in the markets. As a result, these new descriptions, along with the original ones, are displayed to consumers once the app is ready for purchase.

6.2.2 Architecture Overview

Figure 6.4 depicts the workflow of our automated description generation. This takes the following steps:

(1) **Behavior Graph Generation.** Our natural language descriptions are generated via directly interpreting program behavior graphs. To this end, we first perform

Fig. 6.3 Deployment of DESCRIBEME

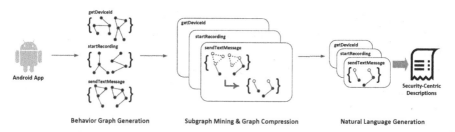

Fig. 6.4 Overview of DESCRIBEME

static program analyses to extract behavior graphs from Android bytecode programs. Our program analyses enable a condition analysis to reveal the triggering conditions of critical operations, provide entry point discovery to better understand the API calling contexts, and leverage both forward and backward dataflow analyses to explore API dependencies and uncover constant parameters. The result of these analyses is expressed via *Security Behavior Graphs* that expose security-related behaviors of Android apps.

(2) **Subgraph Mining and Graph Compression.** Due to the complexity of object-oriented, event-driven Android programs, static program analyses may yield sizable behavior graphs which are extremely challenging for graph traversal and automated interpretation. To address this problem, we next reduce the graph size using subgraph mining. More concretely, we first leverage data mining based technique to discover the *frequent subgraphs* that bear specific behavior patterns. Then, we compress the original graphs by substituting the identified subgraphs with single graph nodes.

(3) **Natural Language Generation.** Finally, we utilize *natural language generation* technique to automatically convert the semantic-rich graphs to human understandable scripts. Given a compressed behavior graph, we traverse all of its paths and translate each graph node into a corresponding natural language sentence. To avoid redundancy, we perform sentence aggregation to organically

combine the produced texts of the same path, and further assemble only the distinctive descriptions among all the paths. Hence, we generate descriptive scripts for every individual behavior graph derived from an app and eventually develop the full description for the app.

6.3 Security Behavior Graph

6.3.1 Formal Definition

Similar to our approach to defining API dependencies in Chap. 3, we consider four factors as essential when describing the security-centric behaviors of an Android app sample: (1) **API call and dependencies**; (2) **Trigger condition**; (3) **Entry point**; and (4) **Constant**.

To address all of the aforementioned factors, we describe app behaviors using *Security Behavior Graphs* (SBG). At a high level, an SBG consists of behavioral operations where some operations have data dependencies.

Definition 1. A *Security Behavior Graph* is a directed graph $G = (V, E, \alpha)$ over a set of operations Σ, where:

- The set of vertices V corresponds to the behavioral operations (i.e., APIs or behavior patterns) in Σ;
- The set of edges $E \subseteq V \times V$ corresponds to the *data dependencies* between operations;
- The labeling function $\alpha : V \to \Sigma$ associates nodes with the labels of corresponding semantic-level operations, where each label is comprised of 4 elements: behavior name, entry point, constant parameter set and precondition list.

Notice that the behavior name can be either an API prototype or a behavior pattern ID. However, when we build the SBGs using static program analysis, we only extract API-level dependency graphs (i.e., the raw SBGs). Then, we perform frequent subgraph mining to identify common behavior patterns so that we can replace the subgraphs with pattern nodes. Graph mining and compression will be discussed in Sect. 6.4.

6.3.2 SBG *of Motivating Example*

Figure 6.5 presents an SBG of the motivating example. It shows that the app first obtains the user's phone number (getLine1Number()) and service provider name (getSimOperatorName()), then encodes the data into a format string (format (String,byte[])), and finally sends the data to network (write(byte[])).

Fig. 6.5 An example SBG

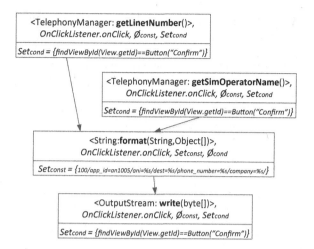

All APIs in this graph are called after the user has clicked a GUI component, so they share the same entry point, OnClickListener.onClick. This indicates that these API calls will be directly triggered by the user.

The sensitive APIs, including getLine1Number(), getSimOperatorName() and write(byte[]), are predominated by a UI-related condition. It checks whether the clicked component is a Button object of a specific name. There exist two security implications behind this information: (1) the app is usually safe to use, without leaking the user's phone number; (2) a user should be cautious when she is about to click this specific button, because the subsequent actions can directly cause privacy leakage.

The encoding operation, format(String,byte[]), takes a constant format string as the parameter. Such a string will later be used to compose the target URL, so it is an important piece of knowledge, that can be used to understand the scenario in which the privacy-related data is used.

6.3.3 Graph Generation

We apply the same techniques illustrated in Chap. 3 to extract API data dependencies, constant parameters and entry points.

Condition Reconstruction We then perform both control-flow and dataflow analyses to uncover the triggering conditions of sensitive APIs. All conditions, in general, play an essential role in security analysis. However, we are only interested in certain trigger conditions for our work. This is because our goal is to generate human understandable descriptions for end users. This implies that an end user should

---

Algorithm 2 Condition Extraction for Sensitive APIs

$SG \leftarrow$ Supergraph
$Set_{<a,c>} \leftarrow$ null
$Set_{api} \leftarrow$ {sensitive API statements in the SG}
for $api \in Set_{api}$ **do**
    $Set_{pred} \leftarrow$ GetConditionalPredecessors(SG,api)
    for $pred \in Set_{pred}$ **do**
        for $\forall var$ defined and used in $pred$ **do**
            $DDG \leftarrow$ BackwardDataflowAnalysis(var)
            $Set_{cond} \leftarrow$ ExtractCondition(DDG, var)
            $Set_{<a,c>} \leftarrow Set_{<a,c>} \cup \{< api, Set_{cond} >\}$
        end for
    end for
end for
output $Set_{<a,c>}$ as a set of $< API, conditions >$ pairs

---

be able to naturally evaluate the produced descriptions, including any condition information. As a result, it is pointless if we generate a condition that cannot be directly observed by a user.

Consequently, our analysis is only focused on three major types of conditions that users can directly observe. (1) *User Interface.* An end user actively communicates with the user interface of an app, and therefore she directly notices the UI-related conditions, such as a click on a specific button. (2) *Device status.* Similarly, a user can also notice the current phone status, such as WIFI on/off, screen locked/unlocked, speakerphone on/off, etc. (3) *Natural environment.* A user is aware of environmental factors that can impact the device's behavior, including the current time and geolocation.

The algorithm for condition extraction is presented in Algorithm 2. This algorithm accepts a supergraph SG as the input and produces $Set_{<a,c>}$ as the output. The supergraph SG is derived from callgraph and control-flow analyses; $Set_{<a,c>}$ is a set of $< a, c >$ pairs, each of which is a mapping between a sensitive API and its conditions.

Given the supergraph SG, our algorithm first identifies all the sensitive API statements, Set_{api}, on the graph. Then, it discovers the conditional predecessors Set_{pred} (e.g., IF statement) for each API statement via GetConditionalPredecessors(). Conditional predecessor means that it is a predominator of that API statement but the API statement is not its postdominator. Intuitively, it means the occurrence of that API statement is indeed conditional and depends on the predicate within that predecessor. Next, for every conditional statement *pred* in Set_{pred}, it performs backward dataflow analysis on all the variables defined or used in its predicate. The result of BackwardDataflowAnalysis() is a data dependency graph DDG, which represents the dataflow from the variable definitions to the conditional statement. The algorithm further calls ExtractCondition(), which traverses this DDG and extracts the conditions Set_{cond} for the corresponding *api* statement. In the end, the API/conditions pair $< api, Set_{cond} >$ is merged to output set $Set_{<a,c>}$.

We reiterate that `ExtractCondition()` only focuses on three types of conditions: user interface, device status and natural environment. It determines the condition types by examining the API calls that occur in the *DDG*. For instance, an API call to `findViewById()` indicates the condition is associated with GUI components. The APIs retrieving phone states (e.g., `isWifiEnabled()`, `isSpeakerphoneOn()`) are clues to identify phone status related conditions. Similarly, if the *DDG* involves time- or location-related APIs (e.g., `getHours()`, `getLatitude()`), the condition is corresponding to natural environment.

User Interface Analysis in Android Apps We take special considerations when extracting UI-related conditions. Once we discover such a condition, we expect to know exactly which GUI component it corresponds to and what text this GUI actually displays to the users.

In order to retrieve GUI information, we perform an analysis on the Android resource files for the app. Our UI resource analysis is different from prior work [12] in two aspects. Firstly, the prior work aims at connecting texts to program entry points, whereas we associate textual resources to conditional statements. Secondly, the previous study did not consider those GUI callbacks that are registered in XML layout files. In contrast, we handle both programmatically and statically registered callbacks in order to guarantee the completeness.

Figure 6.6 illustrates how we perform UI analysis. This analysis takes four steps. First, we analyze the `res/values/public.xml` file to retrieve the mapping between the GUI ID and GUI name. Then, we examine the `res/values/strings.xml` file to extract the string names and corresponding string values. Next, we recursively check all layout files in the `res/layout/` directory to fetch the mapping of GUI type, GUI name and string name. At last, all the information is combined to generate a set of 3-tuples {GUI type, GUI ID, string value}, which is queried by `ExtractCondition()` to resolve UI-related conditions.

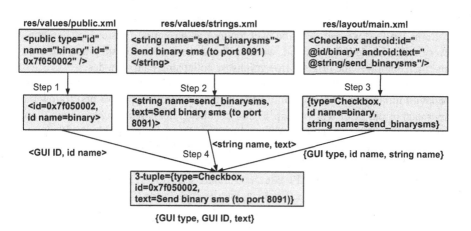

Fig. 6.6 UI resource analysis

Condition Solving Intuitively, we could use a constraint solver to compute predicates and extract concrete conditions. However, we argue that this technique is not suitable for our problem. Despite its accuracy, a constraint solver may sometimes generate excessively sophisticated predicates. It is therefore extremely hard to describe such complex conditions to end users in a human readable manner. As a result, we instead focus on simple conditions, such as equations or negations, because their semantics can be easily expressed using natural language.

Therefore, once we have extracted the definitions of condition variables, we further analyze the equation and negation operations to compute the condition predicates. To this end, we analyze how the variables are evaluated in conditional statements. Assume such a statement is if(hour == 8). In its predicate (hour == 8), we record the constant value 8 and search backward for the definition of variable hour. If the value of hour is received directly from API call getHours(), we know that the condition is current time is equal to 8:00am. For conditions that contain negation, such as a condition like WIFI is NOT enabled, we examine the comparison operation and comparison value in the predicate to retrieve the potential negation information. We also trace back across the entire def-use chain of the condition variables. If there exists a negation operation, we negate the extracted condition.

6.4   Behavior Mining and Graph Compression

Static analysis sometimes results in huge behavior graphs. To address this problem, we identify higher-level behavior patterns from the raw SBGs so as to compress the raw graphs and produce more concise descriptions.

Experience tells us certain APIs are typically used together to achieve particular functions. For example, SMSManager.getDefault() always happens before SMSManager.sendTextMessage(). We therefore expect to extract these behavior patterns, so that we can describe each pattern as an entirety instead of depicting every API included. To this end, we first discover the common subgraph patterns, and then compress the original raw graphs by collapsing pattern nodes.

To this end, we focus on 109 security-sensitive APIs and perform "API-oriented" behavior mining on 1000 randomly-collected top Android apps. We first follow the approach of our prior work (Chap. 3) and conduct *concept learning* to obtain these critical APIs. Then, we construct the subset specific to each individual API. In the end, we apply subgraph mining algorithm [13] to each subset.

Figure 6.7 exemplifies our mining process. Specifically, it shows that we discover a behavior pattern for the API getLastKnownLocation(). This pattern involves two other API calls, getLongitude() and getLatitude(). It demonstrates the common practice to retrieve location data in Android programs.

Now that we have identified common subgraphs in the raw SBGs, we can further compress these raw graphs by replacing entire subgraphs with individual nodes. This involves two steps, subgraph isomorphism and subgraph collapse. We utilize

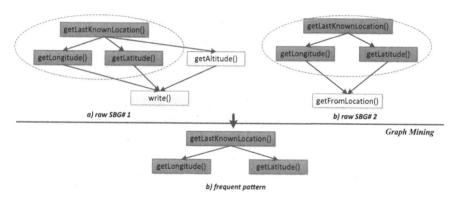

Fig. 6.7 Graph mining for getLastKnownLocation()

the VF2 [14] algorithm to solve the subgraph isomorphism problem. In order to maximize the graph compression rate, we always prioritize a better match (i.e., larger subgraph). To perform subgraph collapse, we first replace subgraph nodes with one single new node. Then, we merge the attributes (i.e., context, conditions and constants) of all the removed nodes, and put the merged label onto the new one.

6.5 Description Generation

6.5.1 Automatically Generated Descriptions

Given a behavior graph SBG, we translate its semantics into natural language descriptions. This descriptive language follows a subset of English grammar, illustrated in Fig. 6.8 using Extended Backus-Naur form (EBNF). The description of an app is a conjunction of individual sentences. An atomic sentence makes a statement and specifies a modifier. Recursively, a non-empty atomic modifier can be an adverb clause of condition, which contains another sentence.

The translation from a SBG to a textual description is then to map the graph components to the counterparts in this reduced language. To be more specific, each vertex of a graph is mapped to a single sentence, where the API or behavioral pattern is represented by a statement; the conditions, contexts and constant parameters are expressed using a modifier. Each edge is then translated to "and" to indicate data dependency.

One sentence may have several modifiers. This reflects the fact that one API call can be triggered in compound conditions and contexts, or a condition/context may accept several parameters. The modifiers are concatenated with "and" or "or" in order to verbalize specific logical relations. A context modifier begins with "once" to show the temporal precedence. A condition modifier starts with either "if" or "depending on if". The former is applied when a condition is statically

Fig. 6.8 An abbreviated
syntax of our descriptions

⟨*description*⟩	::=	⟨*sentence*⟩*
⟨*sentence*⟩	::=	⟨*sentence*⟩ 'and' ⟨*sentence*⟩
	|	⟨*statement*⟩ ⟨*modifier*⟩
⟨*statement*⟩	::=	⟨*subject*⟩ ⟨*verb*⟩ ⟨*object*⟩
⟨*subject*⟩	::=	⟨*noun phrase*⟩
⟨*object*⟩	::=	⟨*noun phrase*⟩
	|	⟨*empty*⟩
⟨*modifier*⟩	::=	⟨*modifier*⟩ ⟨*conj*⟩ ⟨*modifier*⟩
	|	⟨*when*⟩ ⟨*sentence*⟩
	|	⟨*if*⟩ ['not'] ⟨*sentence*⟩
	|	⟨*constant*⟩
	|	⟨*empty*⟩
⟨*conj*⟩	::=	'and'
	|	'or'
⟨*when*⟩	::=	'once'
⟨*if*⟩	::=	'if'
	|	'depending on if'
⟨*empty*⟩	::=	' '

resolvable while the latter is prepared for any other conservative cases. Notice that it is always possible to find more suitable expressions for these conjunctions.

In our motivating example, `getLine1Number()` is triggered under the condition that a specific button is selected. Due to the sophisticated internal computation, we did not extract the exact predicates. To be safe, we conservatively claim that the app retrieves the phone number *depending on if the user selects Button "Confirm"*.

6.5.2 Behavior Description Model

Once we have associated a behavior graph to this grammatical structure, we further need to translate an API operation or a pattern to a proper combination of subject, verb and object. This translation is realized using our *Behavior Description Model*. Conditions and contexts of SBGs are also translated using the same model because they are related to API calls.

We manually create this description model and currently support 306 sensitive APIs and 103 API patterns. Each entry of this model consists of an API or pattern signature and a 3-tuple of natural language words for subject, verb and object. We construct such a model by studying the Android documentation [15]. For instance, the Android API call `createFromPdu(byte[])` programmatically constructs incoming SMS messages from underlying raw Protocol Data Unit (PDU) and hence it is documented as "Create an SmsMessage from a raw PDU" by Google. Our model records its API prototype and assigns texts "the app", "retrieve" and "incoming SMS messages" to the three linguistic components respectively. These three components form a sentence template. Then, constants, concrete conditions and contexts serve as modifiers to complete the template. For example, the template of HttpClient.execute() is represented using words "the app", "send" and "data to network". Suppose an app uses this API to deliver data to a constant URL "http://constant.url", when the phone is locked (i.e., keyguard is on). Then, such constant value and condition will be fed into the template to produce the sentence *"The app sends data to network "http://constant.url" if the phone is locked."* The condition APIs share the same model format. The API checking keyguard status (i.e., KeyguardManager.isKeyguardLocked()) is modeled as words "the phone", "be" and "locked".

It is noteworthy that an alternative approach is to generate this model programmatically. Sridhara et al. [8] proposed to automatically extract descriptive texts for APIs and produce the Software Word Usage Model. The API name, parameter type and return type are examined to extract the linguistic elements. For example, the model of createFromPdu(byte[]) may therefore contain the keywords "create", "from" and "pdu", all derived from the function name. Essentially, we can take the same approach. However, we argue that such a generic model was designed to assist software development and is not the best solution to our problem. An average user may not be knowledgeable enough to understand the low-level technical terms, such as "pdu". In contrast, our text selections (i.e., "the app", "retrieve" and "incoming SMS messages") directly explain the behavior-level meaning.

We generate description model for API patterns based on their internal program logics. Table 6.1 presents the three major logics, we have discovered in behavioral patterns. (1) A singleton object is retrieved for further operations. For example, a SmsManager.getDefault() is always called prior to SmsManager.sendTextMessage() because the former fetches the default SmsManager that the latter needs. We therefore describe only the latter which is associated to a more concrete behavior. (2) Successive APIs constitute a dedicated workflow. For instance, divideMessage() always happens before sendMultipartTextMessage(), since the first provides the second with necessary inputs. In this case, we study the document of each API and describe the complete behavior as an entirety. (3) Hierarchical information is accessed using multiple levels of APIs. For instance, to use location data, one has to first call getLastKnownLocation() to fetch a Location object, and then call getLongitude() and getLatitude() to read the "double"-typed data from this

Table 6.1 Program logics in behavioral patterns

Program logic	How to describe
Singleton retrieval	Describe the latter
Workflow	Describe both
Access to hierarchical data	Describe the former

object. Since the higher level object is already meaningful enough, we hence describe this whole behavior according to only the former API.

In fact, we only create description models for 103 patterns out of the total 109 discovered ones. Some patterns are large and complex, and therefore are hard to summarize. For these patterns, we have to fall back to the safe area and describe them in a API-by-API manner.

In order to guarantee the security-sensitivity and readability of the descriptive texts, we carefully select the words to accommodate the model. To this end, we learn from the experience of prior security studies [2, 3] on app descriptions: (1) The selected vocabulary must be straightforward and stick to the essential API functionalities. As an counterexample, an audio recording behavior can hardly be inferred from the script "*Blow into the mic to extinguish the flame like a real candle*" [2]. This is because it does not explicitly refer to the audio operation. (2) Descriptive texts must be distinguishable for semantically different APIs. Otherwise, poorly-chosen texts may confuse the readers. For instance, an app with description "*You can now turn recordings into ringtones*" in reality only converts previously recorded files to ringtones, but can be mistakenly associated to the permission `android.permission.RECORD_AUDIO` due to the misleading text choice [2, 3].

6.5.3   Behavior Graph Translation

Now that we have defined a target language and prepared a model to verbalize sensitive APIs and patterns, we further would like to translate an entire behavior graph into natural language scripts. Algorithm 3 demonstrates our graph traversal based translation.

This algorithm takes a SBG G and the description model M_{desc} as the inputs and eventually outputs a set of descriptions. The overall idea is to traverse the graph and translate each path. Hence, it first performs a breadth-first search and collects all the paths into Set_{path}. Next, it examines each path in Set_{path} to parse the nodes in sequence. Each node is then parsed to extract the node name, constants, conditions and contexts. The node name *node.name* (API or pattern) is used to query the model M_{desc} and fetch the {subj,vb,obj} of a main clause. The constants, conditions and contexts are organized into the modifier (Cmod) of main clause, respectively. In the end, the main clause is realized by assembling {subj,vb,obj} and the aggregate modifier Cmod. The realized sentence is inserted into the output set Set_{desc} if it is not a redundant one.

Algorithm 3 Generating Descriptions from a SBG

$G \leftarrow \{$A SBG $\}$
$M_{desc} \leftarrow \{$Description model$\}$
$Set_{desc} \leftarrow \emptyset$
$Set_{path} \leftarrow$ BFS(G)
for $path \in Set_{path}$ **do**
　　$desc \leftarrow$ null
　　for $node \in path$ **do**
　　　　$\{$subj,vb,obj$\} \leftarrow$ Query$_{M_{desc}}(node.name)$
　　　　Cmod $\leftarrow$ null
　　　　$Set_{const} \leftarrow$ GetConsts$(node)$
　　　　for $\forall const \in Set_{const}$ **do**
　　　　　　Cmod $\leftarrow$ Aggregate(Cmod,$const$)
　　　　end for
　　　　$Set_{cc} \leftarrow$ GetConditionsAndContext$(node)$
　　　　for $\forall cc \in Set_{cc}$ **do**
　　　　　　$\{$subj,vb,obj$\}_{cc} \leftarrow$ Query$_{M_{desc}}(cc)$
　　　　　　$text_{cc} \leftarrow$ RealizeSentence($\{$subj,vb,obj$\}_{cc}$)
　　　　　　Cmod $\leftarrow$ Aggregate(Cmod,$text_{cc}$)
　　　　end for
　　　　$text \leftarrow$ RealizeSentence($\{$subj,vb,obj,Cmod$\}$)
　　　　$desc \leftarrow$ Aggregate($desc, text$)
　　end for
　　$Set_{desc} \leftarrow Set_{desc} \cup \{desc\}$
end for
output Set_{desc} as the generated description set

6.5.4   Motivating Example

We have implemented the natural language generation using a NLG engine [10] in 3 K LOC. Figure 6.9 illustrates how we step-by-step generate descriptions for the motivating example.

First, we discover two paths in the SBG: (1) getLine1Number() → format() → write() and (2) getSimOperatorName() → format() → write().

Next, we describe every node sequentially on each path. For example, for the first node, the API getLine1Number() is modeled by the 3-tuple {"the app", "retrieve", "your phone number"}; the entry point OnClickListener.onClick is translated using its model {"a GUI component", "be", "clicked"} and is preceded by "Once" ; the condition findViewById(View.getId)==Button("Confirm") is described using the template {"the user", "select", " "}, which accepts the GUI name, *Button "Confirm"*, as a parameter. The condition and main clause are connected using "depending on if".

At last, we aggregate the sentences derived from individual nodes. In this example, all the nodes share the same entry point. Thus, we only keep one copy of *"Once a GUI component is clicked"*. Similarly, the statements on the nodes are

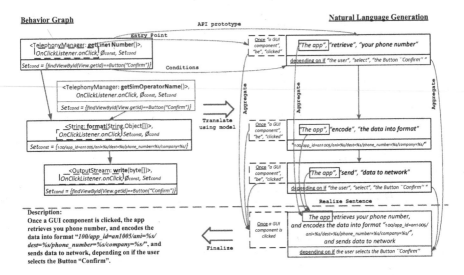

Fig. 6.9 Description generation for the motivating example

also aggregated and thus share the same subject "The app". We also aggregate the conditions in order to avoid the redundancy. As a result, we obtain the description illustrated at the bottom left of Fig. 6.9.

6.6 Evaluation

6.6.1 Correctness and Security-Awareness

Correctness To evaluate the correctness, we produce textual descriptions for DroidBench apps (version 1.1) [11]. DroidBench apps are designed to assess the accuracy of static analyses on Android programs. We use these apps as the ground truths because they are open-sourced programs with clear semantics. Table 6.2 presents the experimental results, which show that DESCRIBEME achieves a true positive rate of 85 %.

DESCRIBEME misses behavior descriptions due to three major reasons. (1) Points-to analysis lacks accuracy. We rely on Soot's capability to perform points-to analysis. However, it is not precise enough to handle the instance fields accessed in callback functions. (2) DESCRIBEME does not process exception handler code and therefore loses track of its dataflow. (3) Some reflective calls cannot be statically resolved. Thus, DESCRIBEME fails to extract their semantics.

Table 6.2 Description generation results for DroidBench

Total #	Correct	Missing desc.	False statement
65	55	6	4

DESCRIBEME produces false statements mainly because of two reasons. First, our static analysis is not sensitive to individual array elements. Thus, it generates false descriptions for the apps that intentionally manipulate data in array . Second, again, our points-to analysis is not accurate and may lead to over-approximation.

Despite the incorrect cases, the accuracy of our static analysis is still comparable to that of FlowDroid [16], which is the state-of-the-art static analysis technique for Android apps. Moreover, it is noteworthy that the accuracy of program analysis is not the major focus of this work. Our main contribution lies in the fact that, we combine static analysis with natural language generation so that we can automatically explain program behaviors to end users in human language. To this end, we can in fact apply any advanced analysis tools (e.g., FlowDroid, AmanDroid [17], etc.) to serve our needs.

Permission Fidelity To demonstrate the security-awareness of DESCRIBEME, we use a description vetting tool, AutoCog [3], to evaluate the "permission-fidelity" of descriptions. AutoCog examines the descriptions and permissions of an app to discover their discrepancies. We use it to analyze both the original descriptions and the security-centric ones produced by DESCRIBEME, and assess whether our descriptions can be associated to more permissions that are actually requested.

Unfortunately, AutoCog only supports 11 permissions in its current implementation. In particular, it does not handle some crucial permissions that are related to information stealing (e.g., phone number, device identifier, service provider, etc.), sending and receiving text messages, network I/O and critical system-level behaviors (e.g., KILL_BACKGROUND_PROCESSES). The limitation of AutoCog in fact brings difficulties to our evaluation: if generated descriptions are associated to these unsupported permissions, AutoCog fails to recognize them and thus cannot conduct equitable assessment. Such a shortcoming is also shared by another NLP-based (i.e., natural language processing) vetting tool, WHYPER [2], which focuses on even fewer (3) permissions. This implies that it is a major challenge for NLP-based approaches to achieve high permission coverage, probably because it is hard to correlate texts to semantically obscure permissions (e.g., READ_PHONE_STATE). In contrast, our approach does not suffer from this limitation because API calls are clearly associated to permissions [18].

Despite the difficulties, we manage to collect 30 benign apps from Google play and 20 malware samples from Malware Genome Project [19], whose permissions are supported by AutoCog. We run DESCRIBEME to create the security-centric descriptions and present both the original and generated ones to AutoCog. However, we notice that AutoCog sometimes cannot recognize certain words that have strong security implications. For example, DESCRIBEME uses "*geographic location*" to describe the permissions ACCESS_COARSE_LOCATION and ACCESS_FINE_LOCATION. Yet, AutoCog cannot associate this phrase to any of the permissions.

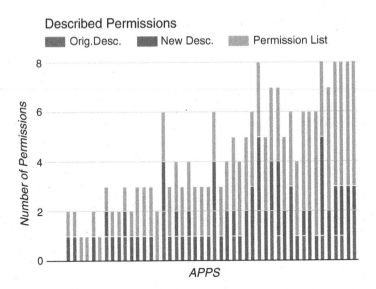

Fig. 6.10 Permissions reflected in descriptions

The fundamental reason is that AutoCog and DESCRIBEME use different glossaries. AutoCog performs machine learning on a particular set of apps and extracts the permission-related glossary from these existing descriptions. In contrast, We manually select descriptive words for each sensitive API, using domain knowledge.

To bridge this gap, we enhance AutoCog to recognize the manually chosen keywords. The experimental result is illustrated in Fig. 6.10, where X-axis represents the evaluated apps and Y-axis is the amount of permissions. The three bars, from top to bottom, represent the amounts of permissions that are requested by the apps, recognized by AutoCog from security-centric descriptions and identified from original descriptions, respectively. Cumulatively, 118 permissions are requested by these 50 apps. Twenty permissions are discovered from the old descriptions, while 66 are uncovered from our scripts. This reveals that DESCRIBEME can produce descriptions that are more security-sensitive than the original ones.

DESCRIBEME fails to describe certain permission requests due to three reasons. First, some permissions are used for native code or reflections that cannot be resolved. Second, a few permissions are not associated to API calls (e.g., RECEIVE_BOOT_COMPLETED), and thus are not included into the SBGs. Last, some permissions are correlated to certain API parameters. For instance, the query API requires permission READ_CONTACTS only if the target URI is the Contacts database. Thus, if the parameter value cannot be extracted statically, such a behavior will not be described.

6.6.2 Readability and Effectiveness

To evaluate the readability and effectiveness of generated descriptions, we perform a user study on the Amazon's Mechanical Turk (MTurk) [20] platform. The goal is two-fold. First, we hope to know whether the generated scripts are readable to average audience. Second, we expect to see whether our descriptions can actually help users avoid risky apps. To this end, we follow Felt et al.'s approach [21], which also designs experiments to understand the impact of text-based protection mechanisms.

Methodology We produce the security-centric descriptions for Android apps using DESCRIBEME and measure user reaction to the old descriptions (Condition 1.1, 2.1–2.3), machine-generated ones (Condition 2.1) and the *new descriptions* (Condition 2.4–2.6). Notice that the *new description* is the old one plus the generated one.

Dataset Due to the efficiency consideration, we perform the user study based on the descriptions of 100 apps. We choose these 100 apps in a mostly random manner but we also consider the distribution of app behaviors. In particular, 40 apps are malware and the others are benign. We manually inspect the 60 benign ones and further put them into two categories: 16 privacy-breaching apps and 44 completely clean ones.

Participants Recruitment We recruit participants directly from MTurk and we require participants to be smartphone users. Besides, we ask screening questions to make sure participants understand basic smartphone terms, such as "Contacts" or "GPS location".

Hypotheses and Conditions *Hypothesis 1: Machine-generated descriptions are readable to average smartphone users.* To assess the readability, we prepare both the old descriptions (Condition 1.1) and generated ones (Condition 1.2) of the same apps. We would like to evaluate machine-generated descriptive texts via comparison.

Hypothesis 2: Security-centric descriptions can help reduce the downloading of risky apps. To test the impact of the security-centric descriptions, we present both the old and *new* (i.e., old + generated) descriptions for malware (Condition 2.1 and 2.4), benign apps that leak privacy (Condition 2.2 and 2.5) and benign apps without privacy violations (Condition 2.3 and 2.6). We expect to assess the app download rates on different conditions.

Study Deployment We post all the descriptions on MTurk and anonymize their sources. We inform the participants that the tasks are about Android app descriptions and we pay 0.3 dollars for each task.

Participants are asked to take part in two sets of experiments. First, they are given a random mixture of original and machine-generated descriptions, and are asked to provide a rating for each script with respect to its readability. The rating is ranged from 1 to 5, where 1 means completely unreadable and 5 means highly readable.

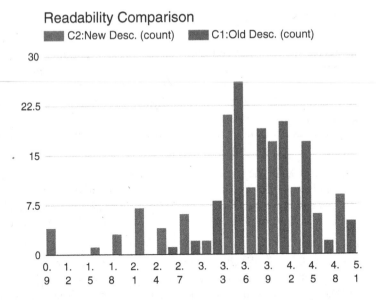

Fig. 6.11 Readability ratings

Second, we present the participants another random sequence of descriptions. Such a sequence contains both the old and *new* descriptions for the same apps. Again, we stress that the *new description* is the old one plus the generated one. Then, we ask participants the following question: "*Will you download an app based on the given description and the security concern it may bring to you?*". We emphasize "security concern" here and we hope participants should not accept or reject an app due to the considerations (e.g., functionalities, personal interests) other than security risks.

Results and Implications Eventually, we receive 573 responses and a total of 2865 ratings. Figure 6.11 shows the distribution of readability ratings of 100 apps for Condition 1.1 and 1.2. For our automatically created descriptions, the average readability rating is 3.596 while over 80 % readers give a rating higher than 3. As a comparison, the average rating of the original ones is 3.788. This indicates our description is readable, even compared to texts created by human developers. The figure also reveals that the readability of human descriptions are relatively stable while machine-generated ones sometimes bear low ratings. In a further investigation, we notice that our descriptions with low ratings usually include relatively technical terms (e.g., subscriber ID) or lengthy constant string parameters. We believe that this can be further improved during an interactive process. User feedbacks and expert knowledge can help us find out more smooth and user-friendly words and expressions to construct our descriptions. We leave this improvement as a future work.

Table 6.3 App download rates (ADR)

#	Condition	ADR
2.1	Malware w/ old desc.	63.4 %
2.2	Leakage w/ old desc.	80.0 %
2.3	Clean w/ old desc.	71.1 %
2.4	Malware w/ new desc.	24.7 %
2.5	Leakage w/ new desc.	28.2 %
2.6	Clean w/ new desc.	59.3 %

Table 6.3 depicts experimental results for Condition 2.1–2.6. It demonstrates the security impact of our new descriptions. We can see a 38.7 % decrease of application download rate (ADR) for malware, when the new descriptions instead of old ones are presented to the participants. We believe that this is because malware authors deliberately provide fake descriptions to avoid alerting victims, while our descriptions can inform users of the real risks. Similar results are also observed for privacy-breaching benign apps, whose original descriptions are not focused on the security and privacy aspects. On the contrary, our descriptions have much less impact on the ADR of clean apps. Nevertheless, they still raise false alarms for 11.8 % participants. We notice that these false alarms result from descriptions of legitimate but sensitive functionalities, such as accessing and sending location data in social apps. A possible solution to this problem is to leverage the "peer voting" mechanism from prior work [22] to identify and thus avoid documenting the typical benign app behaviors.

References

1. Felt AP, Ha E, Egelman S, Haney A, Chin E, Wagner D (2012) Android permissions: user attention, comprehension, and behavior. In: Proceedings of the eighth symposium on usable privacy and security (SOUPS'12), 2012
2. Pandita R, Xiao X, Yang W, Enck W, Xie T (2013) WHYPER: towards automating risk assessment of mobile applications. In: Proceedings of the 22nd USENIX conference on security, August 2013
3. Qu Z, Rastogi V, Zhang X, Chen Y, Zhu T, Chen Z (2014) Autocog: measuring the description-to-permission fidelity in Android applications. In: Proceedings of the 21st conference on computer and communications security (CCS), 2014
4. Sridhara G, Hill E, Muppaneni D, Pollock L, Vijay-Shanker K (2010) Towards automatically generating summary comments for Java methods. In: Proceedings of the IEEE/ACM international conference on automated software engineering (ASE'10), 2010
5. Sridhara G, Pollock L, Vijay-Shanker K (2011) Generating parameter comments and integrating with method summaries. In: Proceedings of the 2011 IEEE 19th international conference on program comprehension (ICPC'11), 2011
6. Moreno L, Aponte J, Sridhara G, Marcus A, Pollock L, Vijay-Shanker K (2013) Automatic generation of natural language summaries for Java classes. In: Proceedings of the 2013 IEEE 21th international conference on program comprehension (ICPC'13), 2013
7. Buse RP, Weimer WR (2010) Automatically documenting program changes. In: Proceedings of the IEEE/ACM international conference on automated software engineering (ASE'10), 2010

8. Sridhara G, Pollock L, Vijay-Shanker K (2011) Automatically detecting and describing high level actions within methods. In: Proceedings of the 33rd international conference on software engineering (ICSE'11), 2011
9. Soot: A Java Optimization Framework (2016) http://www.sable.mcgill.ca/soot
10. SimpleNLG: Java API for Natural Language Generation (2016) https://code.google.com/p/simplenlg/
11. Droidbench-benchmarks (2016) http://sseblog.ec-spride.de/tools/droidbench/
12. Huang J, Zhang X, Tan L, Wang P, Liang B (2014) AsDroid: detecting stealthy behaviors in Android applications by user interface and program behavior contradiction. In: Proceedings of the 36th international conference on software engineering (ICSE'14), 2014
13. Yan X, Han J (2002) gSpan: graph-based substructure pattern mining. In: Proceedings of IEEE international conference on data mining(ICDM'03), 2002
14. Cordella LP, Foggia P, Sansone C, Vento M (2004) A (Sub) graph isomorphism algorithm for matching large graphs. In: IEEE transactions on pattern analysis and machine intelligence, vol 26(10), 2004, pp 1367–1372
15. Reference - Android Developers (2016) http://developer.android.com/reference/packages.html
16. Arzt S, Rasthofer S, Fritz C, Bodden E, Bartel A, Klein J, Traon YL, Octeau D, McDaniel P (2014) FlowDroid: precise context, flow, field, object-sensitive and lifecycle-aware taint analysis for Android apps. In: Proceedings of the 35th ACM SIGPLAN conference on programming language design and implementation (PLDI '14), June 2014
17. Wei F, Roy S, Ou X, Robby X (2014) Amandroid: a precise and general inter-component data flow analysis framework for security vetting of Android apps. In: Proceedings of the 21th ACM conference on computer and communications security (CCS'14), Scottsdale, AZ, November 2014
18. Au KWY, Zhou YF, Huang Z, Lie D (2012) PScout: analyzing the Android permission specification. In: Proceedings of the 2012 ACM conference on computer and communications security (CCS'12), October 2012
19. Android Malware Genome Project (2012) http://www.malgenomeproject.org/
20. Amazon Mechanical Turk (2016) https://www.mturk.com/mturk/welcome
21. Felt AP, Reeder RW, Almuhimedi H, Consolvo S (2014) Experimenting at scale with google chrome's SSL warning. In: Proceedings of the SIGCHI conference on human factors in computing systems, 2014
22. Lu K, Li Z, Kemerlis V, Wu Z, Lu L, Zheng C, Qian Z, Lee W, Jiang G (2015) Checking more and alerting less: detecting privacy leakages via enhanced data-flow analysis and peer voting. In: Proceedings of the 22th annual network and distributed system security symposium (NDSS'15), 2015

Chapter 7
Limitation and Future Work

Abstract In this chapter, we discuss the limitation of our work and propose further improvement as future work.

7.1 Android Malware Classification

Native Code and HTML5-based Apps We perform static analysis on Dalvik bytecode to generate the behavior graphs. In general, bytecode-level static program analysis cannot handle native code or HTML5-based applications. This is because neither the ARM binary running on the underlying Linux nor the JavaScript code executed in WebView are visible from a bytecode perspective. Therefore, an alternative mechanism is necessary to defeat malware hidden from the Dalvik bytecode.

Evasion Learning-based detection is subject to poisoning attacks. To confuse a training system, an adversary can poison the benign dataset by introducing clean apps bearing malicious features. For example, she can inject harmless code intensively making sensitive API calls that are rarely observed in clean apps. Once such samples are accepted by the benign dataset, these APIs are therefore no longer the distinctive features to detect related malware instances. However, our detectors are slightly different from prior works. First of all, the features are associated with behavior graphs, rather than individual APIs. Therefore, it is much harder for an attacker to engineer confusing samples at the behavioral-level. Second, our anomaly detection serves as a sanitizer for new benign samples. Any abnormal behavior will be detected, and the developer is requested to provide justifications for the anomalies. On the other hand, in theory, it is possible for adversaries to launch mimicry attacks and embed malicious code into seemingly benign graphs to evade our detection mechanism. This, by itself, is an interesting research topic and deserves serious consideration. Nevertheless, we note that it is non-trivial to evade detections based upon high-level program semantics, and automating such evasion attacks does not appear to be an easy task. In contrast, existing low-level transformation attacks can be easily automated to generate many malware variants to bypass the AV scanners. DroidSIFT certainly defeats such evasion attempts.

© The Author(s) 2016
M. Zhang, H. Yin, *Android Application Security*, SpringerBriefs in Computer Science, DOI 10.1007/978-3-319-47812-8_7

7.2 Automated Vulnerability Patching

Soundness of Patch Generation The soundness of our approach results from that of slice computation, patch statement placement and patch optimizations. (1) We perform standard static dataflow analysis to compute taint slices. Static analysis, especially on event-driven, object-oriented and asynchronous programs, is known to introduce false positives. However, such false positives can be verified and mitigated during runtime, with our devised shadowing mechanism and inserted patch code. (2) Our patch statement placement follows the standard taint tracking techniques, which may introduce imprecision. Specifically, our taint policy follows that of TaintDroid. While effective and efficient in a sense, this multi-level tainting is not perfectly accurate in some cases. For instance, one entire file is associated with a single taint. Thus, once a tainted object is saved to a file, the whole file becomes tainted causing over-tainting. Other aggregate structures, such as array, share the same limitation. It is worth noting that improvement of tainting precision is possible. More complex shadowing mechanism (e.g., shadow file, shadow array, etc.) can be devised to precisely track taint propagation in aggregations. However, these mechanisms are more expensive considering runtime cost and memory consumption. (3) Our optimizations take the same algorithms used in compilers, such as constant propagation, dead code elimination. Thus, by design, original program semantics is still preserved when patch optimization is applied. (4) In spite of the fact that our approach may cause false positives in theory, we did not observe such cases in practice. Most vulnerable apps do not exercise sophisticated data transfer for Intent propagation, and thus it is safe to patch them with our technique.

Conversion Between Dalvik Bytecode and Jimple IR Our patch statement placement and optimizations are performed at Jimple IR level. So we need to convert Dalvik bytecode program into Jimple IR, and after patching, back to Dalvik bytecode. We use dex2jar [1] to translate Dalvik bytecode into Java bytecode and then use Soot [2] to lift Java bytecode to Jimple IR. This translation process is not always successful. Occasionally we encountered that some applications could not be converted. Enck et al. [3] pointed out several challenges in converting Dalvik bytecode into Java source code, including ambiguous cases for type inference, different layouts of constant pool, sophisticated conversion from register-based to stack-based virtual machine and handling unique structures (e.g., try/finally block) in Java. In comparison, our conversion faces the same, if not less, challenges, because we do not need to lift all the way up to Java source code. We consider these problems to be mainly implementation errors. Indeed, we have identified a few cases that Soot performs overly strict constraint checking. After we patched Soot, the translation problems are greatly reduced. We expect that the conversion failures can be effectively fixed over time. A complementary implementation option is to engineer a Dalvik bytecode analysis and instrumentation framework, so that operations are directly applied on Dalvik bytecode. Since it avoids conversions between different tools, it could introduce minimal conflicts and failures.

Fully Automatic Defense For most vulnerable samples in our experiment, we are able to manually verify the component hijacking vulnerabilities. However, due to the object-oriented nature of Android programs, computed taint slices can sometimes become rather huge and sophisticated. Consequently, we were not able to confirm the exploitable paths for some vulnerable apps with human effort, and thus could not reproduce the expected attack. Developers are faced with the same, if not more, challenges, and thus fail to come up with a solution in time. Devising a fully automated mechanism is therefore essential to defend this specific complicated vulnerability. In principle, our automatic patching approach can still protect these unconfirmed cases, without knowing the real presence of potential vulnerability. That is to say if a vulnerability does exist, AppSealer will disable the actual exploitation on the fly. Otherwise, AppSealer does not interrupt the program execution and thus does not affect usability. With automated patching, users do not have to wait until developers fix the problem.

7.3 Context-Aware Privacy Protection

Soundness of Our Bytecode Rewriting Our static analysis, code instrumentation, and optimizations follow the standard program analysis and compiler techniques, which have been proven to be correct in the single threading context. In the multi-threading context, our shadow variables for local variables and function parameters are still safe because they are local to each individual thread, while the soundness of shadow fields depends on whether race condition vulnerability is present in original bytecode programs. In other words, if the accesses to static or instance fields are properly guarded to avoid race condition in the original app, the corresponding operations on shadow fields are also guarded because they are placed in the same code block. However, if the original app does have race condition on certain static or instance fields, the information flow tracking on these fields may be out of sync. We modeled Android APIs for both analysis and instrumentation. We manually generate dedicated taint propagation rules for frequently used APIs and those of significant importance (e.g., security sensitive APIs). Besides, we have general default rules for the rest. It is well-recognized that it is a non-trivial task to build a fairly complete API model, and it is also true that higher coverage of API model may improve the soundness of our rewriting. However, previous study [4] shows that a model of approximately 1000 APIs can already cover 90 % of calls in over 90,000 Android applications. In addition, it is also possible to automatically create a better API model by analyzing and understanding Android framework, and we leave it as our future work.

Tracking Implicit Flow It is well known that sensitive information can propagate in other channels than direct data flow, such as control flow and timing channels. It is extremely challenging to detect and keep track of all these channels. In this work, we do not consider keeping track of implicit flow. This means that a dedicated malicious Android developer is able to evade Capper. This limitation is also shared by other solutions based on taint analysis, such as TaintDroid [5] and AppFence [6]. Serious research in this problem is needed and is complementary to our work.

Java Reflection A study [7] shows that many Android applications make use of Java reflection to call undocumented methods. While in 88.3 % cases, the class names and method names of these reflective calls can be statically resolved, the rest can still cause problems. In our experiment, we seldom encounter this situation, because even though some apps indeed use reflective calls, they are rarely located within the taint propagation slices. That is, these reflective calls in general are not involved in privacy leakage. We could use a conservative function summary, such that all output parameters and the return value are tainted if any of input parameter is tainted, but it might be too conservative. A more elegant solution might be to capture the class name and the method name at runtime and redirect to the corresponding function summary, which enforces more precise propagation logic. We leave this as our future work.

Native Components Android applications sometimes need auxiliary native components to function, while, unfortunately, static bytecode-level analysis is not capable of keeping track of information flow within JNI calls. However, many apps in fact use common native components, which originate from reliable resources and are of widely recognized dataflow behavior. Thus, it is possible to model these known components with offline knowledge. In other words, we could build a database for well-known native libraries and create proper function summaries for JNI calls, and therefore exercise static data propagation through native calls with the help of such summaries.

7.4 Automated Generation of Security-Centric Descriptions

The correctness and accuracy of generated description are largely affected by that of static program analysis. Static dataflow analysis is conservative and may cause over-approximation. We believe more advanced static analysis techniques can help reduce false statements of generated descriptions. Besides, in order not to mislead end users, DESCRIBEME attempts to stay on the safe side and does not over-claim its findings.

Static bytecode-level program analysis may also yield false negatives due to the presence of Java reflections, native code, dynamically loaded classes or JavaScript/HTML5-based code. Fundamentally, we need whole-system dynamic analysis (e.g., DroidScope [8]) to address runtime behaviors. Nevertheless, from static analysis, we can observe the special APIs (e.g., `java.lang.reflect.Method.invoke()`) that trigger those statically unresolvable code. Further, DESCRIBEME can retrieve and describe the dependencies between these special APIs and other explicit and critical ones. Such dependencies can help assess the risks of former. Even if there exists no such dependencies, DESCRIBEME can still directly document the occurrence of special APIs. The prevalence of unresolved operations and the drastic discrepancy between our description and permission requests are clues for end users to raise alert.

References

1. dex2jar (2016) http://code.google.com/p/dex2jar/
2. Soot: A Java Optimization Framework (2016) http://www.sable.mcgill.ca/soot/
3. Enck W, Octeau D, McDaniel P, Chaudhuri S (2011) A study of Android application security. In: Proceedings of the 20th usenix security symposium, August 2011
4. Chen KZ, Johnson N, D'Silva V, Dai S, MacNamara K, Magrino T, Wu EX, Rinard M, Song D (2013) Contextual policy enforcement in Android applications with permission event graphs. In: Proceedings of the 20th annual network and distributed system security symposium (NDSS'13), February 2013
5. Enck W, Gilbert P, Chun BG, Cox LP, Jung J, McDaniel P, Sheth AN (2010) TaintDroid: an information-flow tracking system for realtime privacy monitoring on smartphones. In: Proceedings of the 9th USENIX symposium on operating systems design and implementation (OSDI'10), October 2010
6. Hornyack P, Han S, Jung J, Schechter S, Wetherall D (2011) These aren't the droids you're looking for: retrofitting Android to protect data from imperious applications. In: Proceedings of CCS, 2011
7. Felt AP, Chin E, Hanna S, Song D, Wagner D (2011) Android permissions demystified. In: Proceedings of CCS, 2011
8. Yan LK, Yin H (2012) DroidScope: seamlessly reconstructing OS and Dalvik semantic views for dynamic Android malware analysis. In: Proceedings of the 21st USENIX security symposium, August 2012

Chapter 8
Conclusion

Abstract In this chapter, we conclude our work and summarize this book.

To battle various security threats in Android applications, we propose a semantics and context-aware approach. We argue that such an approach improves the effectiveness Android malware detection and privacy preservation, ameliorates the usability of security-related app descriptions and can solve complex software vulnerabilities in mobile apps. Our argument has been validated via the design, implementation and evaluation of a series of security enhancement techniques.

DroidSIFT demonstrated that semantics-aware Android malware classification not only achieves high detection rate and low false positive and negative rates, but also defeats polymorphic and zero-day malware. Moreover, it is resilient to bytecode-level transformation attacks and outperforms all the existing antivirus detectors with respect to the detection of obfuscated malware.

AppSealer showed that with the analysis of program semantics in advance, the patch of complex application vulnerabilities can be automatically generated. In addition, static program analysis facilitates a selective patch code instrumentation and therefore improves the runtime performance of patched programs.

Capper illustrated that a context-aware privacy policy can effectively differentiate legitimate use of private user data from real privacy leakage, because program context can faithfully reflect the true intention of critical operations.

DESCRIBEME showed that natural language app descriptions, of better readability and higher security sensitivity, are created via program analysis and comprehension of application semantics. Automatically produced descriptions can help users avoid malware and privacy-breaching apps.

© The Author(s) 2016   105
M. Zhang, H. Yin, *Android Application Security*, SpringerBriefs in Computer
Science, DOI 10.1007/978-3-319-47812-8_8

Printed in the United States
By Bookmasters